Frederick Dixon Chester

A Preliminary Arrangement of the Species of the Genus Bacterium

Frederick Dixon Chester

A Preliminary Arrangement of the Species of the Genus Bacterium

ISBN/EAN: 9783337801106

Printed in Europe, USA, Canada, Australia, Japan

Cover: Foto ©berggeist007 / pixelio.de

More available books at **www.hansebooks.com**

A PRELIMINARY ARRANGEMENT OF THE SPECIES OF THE GENUS BACTERUM.

CONTRIBUTIONS TO DETERMINATINE BACTERIOLOGY. PART I.

General Introduction.

When about a year since the writer began a study of the bacterial flora of cultivated soils, an important obstacle was encountered at the outset. Species could be isolated and carefully studied, but to determine whether such species were new, or identical with those already described, was a matter of some difficulty and much uncertainty.

The reason of this, as all bacteriologists will appreciate, lay in the fact that hitherto no attempt had ever been made to arrange the something like 600 species of Schizomycetes unto anything like a system sufficiently minute to enable one to identify a bacterial form with anything approaching certainty. Not only did we lack such a system, but the attempt to create one for ourselves seemed an unsurmountable task. The difficulty in the way of an arrangement or classification of the bacteria lies mainly in the fact that three-fourths of the species already named have been so imperfectly and, perhaps, inaccurately described, that there are but few differential characters to form the basis of a separation or grouping. Another difficulty arises from the fact that the cultural characters of certain bacteria are quite variable, dependent upon media, conditions of growth, etc. This has led past authors to create many unnecessary new species, based upon characters which lie within the limits of normal specific variation.

This consequent multiplication of bad species adds much to the difficulty of arranging them into a satisfactory system.

The subject of systematic bacteriology is in its earliest infancy. The group Schizomycetes, to which the bacteria belong, has not received the same attention from the botanists as other groups of Cryptogams. The opinion has been too generally current that bacterlogy is a sort of adjunct to medical science, and that it belonged to the doctors. The result of this prevailing patronage of the medical men has been the admirable development of the pathological side of the subject, with comparatively little attention paid to its general and biological aspects. I make, therefore, a special appeal for a larger study of the Schizomycetes, not merely as pathogenic agents in some minor relation, but as a group of plants demanding from botanists the same attention as any other group of cryptogams. This means especially the study af bacteriology from a systematic standpoint.

The aim of all botanical investigation is the study of relationships between forms, the aim of the knowledge of relationships is the creation of a scheme of differentiation or classification. Any system of classification is at first imperfect. Specific and even generic distinctions need readjustment. Artificial systems of classification give way to more rational ones. The establishment of a system of classification, convenient at the time, but of a recognized provisional character becomes, however, a base of operations, giving rational direction to biological study, whose final end is the perfection of this original provisional system. In other words, we must arrange biological forms according to some scheme, however crude, and look to future work to perfect or readjust it. This then is my apology for undertaking the work embodied in this paper.

In the following pages an attempt has been made to arrange the known forms of the Genus *Bacterium* into what I can only designate as a purely artificial and therefore imperfect system. The species have been arranged solely with a view to their determination, hence the present scheme, while useful for its purpose, would in some instances be open to grave criticism, should it be supposed to be a natural system of classification as well. For instance, two species of the *Proteus* type might in one case produce a small amount of a yellow pigment, in the other not; they might be closely related species or merely varieties, and hence in a natural system they would stand together, but in an artifical grouping, intended for purposes of identification, they would be more widely separated. Again, it is true, as shown by Marshall Ward,[1] that the same species may show differences in its power to liquefy gelatin, as *B. fluorescenes liquefaciens* and *B. fluorescens non-liquefaciens*, which Ward considers may be merely varieties of the same species. These marked variations if consistent with species differentiation tend to separate more widely an artificial system intended for purposes of identification from a more rational system based upon truer relationships. In short the rather wide variations in specific characters will tend to make these two systems diverge quite widely the one from the other.

The shortcomings in the present attempt are only too apparent to the writer, and hence much criticism is to be expected from bacteriologists. Crude and imperfect as is the present system of arrangement I believe it is better than nothing, and therefore I hope it will be of use to those who are working among the Saprophitic bacteria.

Nomenclature and Descriptions.

The most important need in bacteriology to-day is a definite system of nomenclature by which the descriptions of the author can be accurately and in the fewest possible words conveyed to the reader.

In all branches of biological science, specific forms, as animals, or plants, are accurately described, but back of those descriptions is the original type specimen, kept in some great herbarium or museum, from which the author's description was written and from which a new description can at any time be easily and quickly made.

In bacteriology the majority of the types from which the original authors wrote such crude and imperfect descriptions have been lost.

Once seen by their discoverer they are, perhaps, rediscovered by

---

[1] Proc. Roy. Soc., LXI, 1897. p. 415.

another and redescribed as a new species, hence we are never sure from the imperfect descriptions of older authors, with no type specimens in existence in pure culture to serve as reference, whether newly discovered forms are or are not identical with the older species. Thus in the absence of type specimens of bacteria, always available for reference, it is all the more important that bacterial species should be so minutely and accurately described that there will be no difficulty in the future for bacteriologists to identify species from the written descriptions· It is hoped that the general use of formaline to kill and preserve cultures will enable us to retain for reference cultures on the most important media of all newly-described species of the future.

To accurately describe a species and its cultural characters, terms having a definite meaning should be employed. By this means a single designating word will convey a better and more accurate meaning than a whole sentence of description after the author's own peculiar style of expression. Descriptions of cultures should be brief and pointed. Such descriptions cannot be made without the use of clearly-defined and well understood terms; in other words, a system of bacteriological nomenclature is a very fundamental and pressing need. In the following table I, the author has attempted to construct a limited system of terms which have been employed throughout the present paper wherever practicable, one of the main objects in view being to condense the specific descriptions which follow; to further aid in this matter abbreviations have been also largely used, the meanings of which will be found in an accompanying table.

It is an open question among bacteriologists as to what should constitute a complete description of a bacterial species, also as regards the number of media to be employed. On general principles it would appear that the more minute and thorough the descriptions of the cultural characters in the maximum number of media, the more satisfactory the result. There is danger, however, that this thoroughness may be carried too far, leading us to make specific distinctions based upon minute and really trivial differences in habit or behavior.

Until bacteriologists can universally adopt uniform formulæ for making media, with closely approximate alkalinity, acidity or neutrality, so that a given medium in one laboratory will practically be identical with the same from another laboratory we cannot expect satisfactory results. It is very important that the bacteriologist should study his forms for a sufficient length of time, and under such slightly varying conditions of media and temperature as will invariably prevail at different times, to be sure that the cultural characters of his described forms may not be too hastily formulated. These difficulties are largely overcome by making the media as absolutely uniform as possible and growing in a cold box kept uniformly between 20°–22° C instead of at varying room temperatures, as well as in the thermostat kept at 37° C.

In the absence of uniformity in many matters of technique, it is best for the bacteriologist to describe (1) how his media are prepared and neutralized, (2) methods of staining, (3) how chemical tests are made, such as indol, phenol, nitrites, etc.

In the following table II the writer has given a scheme for the description of a bacterial species of sufficient detail, it is thought, for most purposes.

TABLE I.
CHARACTERS OF BACTERIA.

I Gelatin Stab Cultures.
  A Non-Liquefying.
  1 *Line of Puncture* (a) *filiform*, uniform growth without any special characters (b) *tuberculate*, papillate, echinulate extensions (c) *villous*, beset with long or short undivided hair like extensions (d) *arborescent*, beset with branched hair like extensions (e) *beaded*, composed of small, round more or less conjointed colonies (f) *banded* longitudinally.
  2 *Surface Growth* (see descriptions of colonies on plate).
  B Liquefying.
  (a) *crateriform*, saucer shaped liquefaction of the gelatin (b) *saccate*, shape of an elongated sack, tubular, cylindrical (c) *funnel formed* (d) *napiform*, outline of a turnip (e) *fusiform*, outline of a parsnip (f) *stratiform*, liquefaction extending to the walls of the tube and then downward horizontally.
II Smear Cultures (see plate culture characters).
III Bouillon Cultures.
  1 *Medium*, clear, turbid.
  2 *Sediment*, floculent, stringy, granular.
  3 *Surface Growth*, membrane (see plate culture characters).
IV Plate Cultures.
  A Non-Liquefying.

1 FORM (a) *punctiform*, dimensions too slight for defining form by naked eye, minute, raised, semispherical (b) *round* (c) *irregular* (d) *oval* (e) *fusiform* (f) *cochleate*, spiral or twisted like a snail shell (g) *ameboid*, very irregular like the changing forms of amœbæ (h) *conglomerate*, an aggregate of colonies of the same or irregular form.

2 *Surface Elevation.*

(a) *flat*, spreading, thin (b) *raised*, growth thick with abrupt terraced edges (c) *convex*, surface the segment of a circle, but very flatly convex (d) *pulvinate*, surface the segment of a circle, but decidely convex (e) *capitate*, semi-spherical (f) *rough*, irregular elevations and depressions (g) *contoured*, like the undulating surface of a relief map (h) *papillate*, horn like projections (i) *rugose*, wrinkled (j) *alveolate*, depressions separated by thin walls (k) *pitted*, (l) *sulcate*, ridged or furrowed.

3 OPTICAL CHARATERS OF THE SURFACE.

(a) *moist glistening* (b) *dull* (c) *greasy* (d) *farinaccous*, like meal, dry, granular (e) *transparent, translucent*, (f) *iridescent*, reflection like that of "mother of pearl" (g) *opaque* (h) *chalky*.

4 CONSISTENCY.

(a) *thin*, (b) *membraneous*, thin, dry, separating from medium (c) *coriaceous*, thick like leather or parchment (d) *viscous*, ropy (e) *slimy* (f) *gelatinous* (g) *brittle*.

5 *Edge of Colony.*

(a) *entire, undulate, repand* (b) *erose*, finely eroded as if gnawed (c) *lobed, auriculate* (d) *laciniate*, cut jaggedly into deep narrow lobes (e) *lacerate*, cut variously into irregular segments (f) *fimbriate*, edge boardered by slender processes thicker than hairs (g) *cilliate* (h) *tufted* (i) *flocose*, wooly, filaments in fleecy masses (j) *curled*, filaments in locks or ringlets (k) *filamentous*, consisting of loosely placed interwoven filaments, not so dense as flocose.

6 Internal Structure of Colony.

(a) *homogeneous*, uniform throughout (b) concentrically zoned (c) *marmorated*, traversed by veins as in some kinds of marble, marbled (d) *finely punctate* (e) *areolate*, marked out with small spaces, reticulate (f) *moruloid*, having the character of a morula, resembling a mulberry, segmented (g) *finely granular* (h) *coarsely granular* (i) *grained*, as in lumber (j) *curled*, composed of twisted bundles of parallel filaments as in locks or ringlets (k) *flocose, filamentous, myceloid*, as above.

TABLE II.

DESCRIPTION OF BACTERIA, SCHEME.

I Name.

II Synonyms.

III Habitat.

VI Morphology (1) motility (2) form, size and groupings, on agar, in bouillon, (3) spores.

V Staining peculiarities (1) in aqueous analine solutions (2) carbol fuchsin, Ehrlich's solution (3) staining by Gram's method (4) flagella demonstration.

VI Temperature conditions of growth (1) minimum, maximum, optimum (2) thermal death point.

VII Gelatin colonies (1) deep colonies (a) naked eye characters; form, color, (b) with AA objective; form, color, edge, internal structure.

(2) Surface colonies (a) naked eye characters; form, color, elevation, (b) with AA objective; form, color surface elevation, optical characters, surface, consistency, edge of colony, internal structure of colony.

VIII Agar colonies. do.

IX Gelatin stab. (A) Non liquefying, (1) line of puncture (2) surface growth (See description of gelatin surface colonies). (B) Liquefying (a). Crateriform, saccate, napiform, fusiform, funnel, stratiform, (b) *Fluid*, clear, turbid, floculent (c) Sediment (+, or, —,), floculent, stringy, granular. (d) Membrane (+ or —), character and color.

X Gelatin slant (in a non liquefying species.) } Inoculation made
XI Agar slant, glycerin agar slant, blood serum. } by one longitudi-
XII Potato culture. } nal medial stroke
} of an infected oese.

(1) limited to track of inoculation or spreading, color, surface elevation, optical character of surface, consistency, edges.

XIII Bouillon cultures. (1) *Fluid*, clear, turbid. (2) *Sediment* floculent, stringy, granular (3) surface growth.

XIV Plain milk. Coagulation, peptonization, saponification, rendered viscous, color, consistency, changes in reaction.

XV Litmus milk. See above. Changes in color; deepened, reddened, decolorized. Changes of reaction tested with litmus paper.

XVI Pepton rosolic acid. solution. See bouillon cultures. Changes of color.

XVII Fermentation tube tests (1) presence or absence of gas production in glucose, lactose, and sucrose bouillon. (2) presence (facultative anaerobic) or absence (aerobic) of growth in closed arm of the fermentation tube.

If gas be produced, absorption with NaOH, and determination of $\frac{H}{CO_2}$ ratio.

XVIII Chemical relations. Iudol, phenol, H_2S, reduction of nitrates to nitrites, production of acid in 2 per cent glucose bouillon by titration with $\frac{N}{10}$ NaOH, after 5 days growth; production of acid or alkalies in nutrient bouillon, after 5 days by titration as above; pigment formation, solubility in water, alcohol, ether, bisulphide of carbon, benzole, chloroform, action of acids and alkalines on the solutions of the pigments.

XIX Pathogenesis.

TABLE OF ABREVIATIONS.

USED IN TEXT.

(+) Has a positive significance, indicates presence in marked quantity or a strong development.

(—) Indicates the absence of a thing or no development.

($\triangle$) Has a diminutive significance, indicates slow development, slight growth, or the presence of a small quantity.

(?) Questionable, or imperfectly known.

(±) Indicates presence or absence, also that the culture medium remains unchanged as M ±. P P ±.

(→) Becoming, eventually, developing into.

A, agar—ac, acid—ae, aerobic—alk, alkaline—an, anaerobic—am, amphoteric—assoc, associated.

B, bacillus, organism, bouillon—Bs, blood serum—bl, blue—br, brown.

Cs, colonies—Co, coagulated, Co—. not coagulated—coal, coalescent—cults, cultures.

D, days, dark—diff, difference, points of differentiation—decol, decolorized.

Elev, elevated, raised growth.

Fac, facultative—fla, flagella—floc, floculent, floculi—fluor, fluorescence.

G, gel. gelatin—gl, glistening—Gl, glucose—glyc, glycerine—g, green—gr, gray—Gm+, stain by Grams method—Gm—, do not stain as above—gran, granular.

Hab, habitat, material from which an organism was isolated—In, indol—inoc, inoculation—inj, injection—inper, injection into the peritoneal cavity—inven, injection into a vein—indt, indeterminate, not distinguishable.

Lac, lactose—l, light, pale—limtd, limited—lit, litmus—liq, liquid.

M, milk—mo, moist—mot, motile, motility as Mot △, memb, membrane—med, medium.

P, potato—pel, pelicle—pep, peptonized—ph, phenol—PR, peptone rosolic acid solution—phos, phosphorescence—pur, purulent.

Rd, red.

Su, Sucrose, cane sugar—sed, sediment—sbc, subcutaneous inoculation—spdg, spreading—sl, slant, smear culture—st, stab culture.

Transl, translucent—transp, transparent.

Wh, white—yl, yellow.

Used in Bibliography.

A.H. Archiv. f. Hygiene.
A. f. D. Archiv. f. Dermatol. u. Syphilis.
A.K.G. Arbeiten aus dem Kaiserl. Gesundheitsamte.
A. Expt. Path. Archiv. f. experim. Pathologie u. Pharmakologie.
B. k. W. Berlin Klin. Wochenschr.
B.B. Beiträge zur Biologicid. Pflansen von F. Cohn.
C. Centralblatt f. Bakteriologie.
C.C. Centralblatt. f. Bakteriologie II. Abtheilg.
Bot. C. Botanisches Centralblatt.
Ch.A. Charité Annalen.
D.M.W. Deutsche mediz. Wochenschrift.

F.M. Fortschritte d Medizin.

B.J. Baumgartens Jahresberichte pathogenen Mikroorganismen.

J.P. Jour Pathol and Bacteriology.

M.M.W. Münch. med. Wochenscrift.

M.K.G Mitteilungen a. d. Kaiserl. Gesundheitsamte.

M. Text books, compendiums of various authors.

N.P. Die Pflanzenfamilien. Migula on Schizomycetes.

A.P. Annales de l'Institut Pasteur.

HyR. Hygien Rundschau.

Ri. M. Riforma medica.

S.M. Semaine médicale.

V.A. Virchows Archiv.

W.M.W. Wiener, med. Wochenschrift.

W.K. W Wiener, klinische Wochenschrift.

Z.H. Zeitsch. f. Hygiene.

Z.K.M. Zeitsch. f. klin. Medizin.

CLASSIFICATION OF THE SCHIZOMYCETES. BACTERIA.[1]

SCHIZOMYCETES.

Very small, one celled chlorophyl free, colorless—rarely weakly red or green colored organisms, which divide in one, two or three directions of space, and which are arranged in a thread like, plate like or cubical manner. The outer membrane of an albuminous nature. Cell contents generally homogenous without a cell nucleus or at most a so-callled central body (*centralkörper*). Sexual reproduction wanting. Resting cells in the form of *endospores* or *gonidia*.

I. Cells in their free condition globular, become but slightly elongated before division. Cell division in 1, 2 or 3 directions of space, Order—COCCACEÆ.

A. Cells without Flagella.

1. Division only in one direction of space.........*Streptoccocus* (Billroth)

2. Division in two directions of space...............*Micrococcus* (Hallier)

3. Divisions in three directions of space................*Sarcina* (Goodsir)

B. Cells with Flagella·

[1]The following classification of the Schizomycetes has been taken from Migula, Lie-ferung 128; Die Naturlichen Pflanzenfamilien, 1896. Except for the Order Bacteriaceæ, since Migulas distinctions cannot be applied to the great mass of imperfectly described members of this order.

1. Divisions in two directions of space.............*Planococcus* (Migula)
2. Divisions in three directions of spaces..............*Planosarcina* (Migula)
II. Cells short or long, cylindrical in form, straight, curved or spiral. Before division increasing to double their length, without a sheath surrounding the chains of individuals, motile or non-motile, endospores present or absent. No true branching.

A. Cells longer than broad, generally 2–6 times, straight or only with an angular bend, never curved or spiral, division only at right angles to axis of rod ; with or without flagella and endospores.

Order. BACTRIACEÆ.

1. Without Endospores...............*Bacterium*.
2. With Endospores...............*Bacillus*.

B. Organisms curved or spirally bent, generally motile through polar flagella. Order SPIRILLACEÆ.
1. Cells stiff, not flexile.
    a. Cells without flagella.........*Spirosoma* (Migula)
    b. Cells with flagella.
       * Cells with one, very rarely with two polar flagella.
            Microspira (Schröter)
       ** Cells with a bundle of polar flagella.
            Spirillum (Ehrenb).
2. Cells flexile..............................*Spirochata* (Ehrenb)
III. Cells short or long cylindrical or clavate-cuneate in form without a sheath surrounding the chains of individuals. Without endospores, with possibly the formation of gonidia like bodies. With true dichotomous branching (this forms an important group standing between the true Bacteriaceae, and the true fungi on the one hand Chlamydobacteriaceæ on the other).

Order MYCOBACTERIACEÆ (Chester).

A. In cultures possessing the characters of true bacteria. Growth on sold media smooth, flat, spreading. Rods with swoolen ends, or cuneate or clavate forms.

Corynebacterium (Lehm-Neum)

B. Cultures on solid media raised, folded or warty. Generally short slender rods, rarely short branched. Take the tubercle stain.

Mycobacterium (Lehm-Neum)

IV. Thread like composed of individual cells, surrounded by a sheath. Simple or with true branching. Ordinary vegetative growth by division in only one direction of space *i. e.* at right angles to the longer axis.

A. Cell contents without sulphur granules.

Order CHLAMYDOBACTERIACEAE.

1. Filaments unbranched.
   a. Cell division only in one direction of space.
   b. Cell division in gonidia formation in three directions of space..........................*Streptothrix* (Cohn)
      * Cells surrounded by a very delicate, hardly discernible sheath (marine)..............*Phragmidiothrix* (Engler)
      ** Sheath easily discernible (fresh water).

Crenothrix (Cohn)

2. Filaments branched...................... ....*Cladothrix* (Cohn).

B. Cell contents with sulphur granules.

Thiothrix (Winogradsky).

IV. Thread like, without a capsule, but with an undulating membrane, and as in Osillaria motile cell contents show sulphur granules. Formation of condia not certainly known.

Order BEGGIATOACEÆ.

A. Threads apparently not septated, septae only faintly visible with iodine staining. Colorless or faintly rose colored.

Beggiatoa (Travisan).

SYNOPSIS OF GENUS BACTERIUM.

I. Aerobic and Facultative Anaerobic.
   A. Without pigment on gelatin or agar.
      1. Grow only at body temperatures.
         a. Grow only on Blood Serum or especially prepared media.
            Non-Motile—Class I ; p. 66.
         b. Grow very poorly, or scarcely at all, at room temperatures.
            Class I A ; p. 67.
      2. Grow at room temperatures, 20°—24° C.
         a. Grow in nutrient gelatin or some of the ordinary culture media.
            * Colonies on gelatin plates, roundish, not ameboid or Proteus like.
            Gelatin not liquefied.
            Decolorized by Gram's Method.
            Motile—Class II ; p. 67.
            Non-Motile—Class III ; p. 77.

Stained by Gram's Method.
Motile—Class IV ; p. 87.
Non-Motile—Class V ; p. 88.
Gelatin liquefied.
Motile—Class VI ; p. 89.
Non-Motile—Class VII ; p. 98.
** Colonies on gelatin plates, becoming streaming, forked,
ameboid, twisted, irregular, cochleate.
Gelatin liquefied—Motile.
Stained by Gram's Method—Class VIII ; p. 101.
Not stained by Gram's Method—Class IX ; p. 102.
Gelatin not liquefied—Class X ; p. 103.
B. Produce pigment on gelatin or agar [Chromogenic Bacteria].
1. Pigment yellowish on gelatin.
a. Gelatin liquefied.
* Motile—Class XI ; p. 104.
** Non-Motile—Class XII ; p. 106.
b. Gelatin not liquefied.
* Motile—Class XIII ; p. 109.
** Non-Motile—Class XIV ; p. 110.
2. Pigment reddish on gelatin or agar.
a. Gelatin liquefied.
* Motile—Class XV ; p. 112.
** Non-Motile—Class XVI ; p. 113.
b. Gelatin not liquefied—Class XVII ; p. 115.
3. Pigment, brownish, black, gray on gelatin—Class XVIII; p. 116.
4. Pigment violet, blue on gelatin and agar—Class XIX ; p. 117.
5. Pigment greenish—Class XX ; p. 118.
C. Colonies, colorless or colored slightly yellowish or greenish, but
with a yellow green or blue green fluorescence.
1. Gelatin liquefied.   [Fluorescent Bacteria.]
a. Motile—Class XXI ; p. 118.
b. Non-Motile—Class XXII ; p. 121.
2. Gelatin not liquefied.
a. Motile—Class XXIII ; p. 121.
b. Non-Motile, Class XXIV ; p. 124.
II. Anaerobic (obligative)—Class XXV ; p. 125.

CLASS I. GROW BEST ON BLOOD, BLOOD SERUM OR ESPECIALLY PREPARED MEDIA, AT BODY TEMPERATURE. NON-MOTILE.

I. Grow only on blood or agar moistened with blood or on especially prepared media.

    1. Bact. influenzae (R. Pfeiffer).

B. 0.2-0.3:0.5, commonly in 2's. Stains with Löffler's alk bl and carbol fuchsin. Gm—. *On A moistened with blood*, in 24-48 hours small glassy drops; older Cs with a ylsh-brsh center.

Nastiukow's Sol* 37°, 24 hrs, small wh flecks at bottom of tube-chains of B.

Nastiukow's Agar* Cs. small gr points (b) round, yl, transl.

Path. Mice imper, ⅓ A sl, death. *Rabbits*, A sl inven, fever, mucular weakness; sbc, ⅙ A sl, knotty thickenings, suppurative.

Diag. (1) Cover glass preparations from bronchial secretions, sputum, etc. (2) Smear cultures on agar moistened with blood; plate cultures with Nastiukow's Agar (C. XVII, 492)*

Hab. Nasal and bronchial secretions, urine, of man affected with influenza.

    2. Bact. pseudo influenzae (Pfeiffer).

Like the preceding, but rods somewhat longer than those of true influenza B. *Path?*

Hab. From broncho pneumonia; Otitis media; Assoc. with influenza.

II. Grow well on blood serum.

    3. Bact. septicus acuminatus (Sternberg).

B. With pointed ends resembling B. of mouse septicæmia, but rather thicker; stains unevenly. Cs small, round, flat, transl→+by coal ylsh layer.

Path. mice—; rabbits—; guineas 2-6 d death, septicæmia.

Hab. Blood and organs of newly-born infant with septicæmia.

    4. Bact. of Lumnitzer.

B. 1.5-2, round ends, △ curved. A sl—*Bs*, 37°, small semisph, grshwh Cs→ coal, with odor of sputum in putrid bronchitis. *Path, mice*, inper. 24 hrs, pur peritonitis, *rabbits*, inj into lung, pneumonia, pleuritis.

Hab. sputum in putrid bronchitis.

    5. Bact. conjunctivitis (Morax).

B. 1.0:2.0, ends rather squared, in 2's—chains; Gm—. *Bs* 24 hrs, moist points, 2d small liq depressions, transp. *Path* smaller animals—man, inflamation of conjunctival sac.

Hab. From chronic inflammation of conjunctiva.

III. Grow only in the presence of pathological secretion, not in blood or blood serum.

6. B. *vaginæ* (Döderlein).

(D derlein's Scheidenbacillus).

B middle sized, rather slender. Grow in Gl B containing 1 per cent. of secretion, can then be transferred to Glyc A=dewy drop like Cs. Fac an. *Path ?* *Hab.* in vaginal secretions.

IV. Cultural character not known.

7. B. ulceris cancrosi (Ducrey).

B. 0.5:1.5, ends rounded, mostly contracted in the middle, in chains. Gm—. Stain ¼ hr. with Löffler's alk. bl, wash but a short time in alchohol. *Hab.* From secretions in soft shanker.

CLASS Ia. *Grows very poorly or scarcely at all at room temperatures. Non-motile.*

8. Bact. conjunctivitis (Koch-Kartulis).

B. 0.25:1.0, in 2's or chains in the pus eells. Gm—; *A sl, B s* 37° isolated Cs-+ confluent, gl, elev. *Gel.* △, liq—. *Path.* on cornea of asses ; dogs, guineas, rabbits negative ; on human conjunctiva positive in one out of six inocs. *Hab.* assoc. with conjunctival catarrh in Egypt.

THE SPECIES WHICH FOLLOW GROW ON ORDINARY NUTRIENT MEDIA.

CLASS II. GELATIN NOT LIQUEFIED. MOTILE.

I. Gas Generated in Glucose Bouillon.

A. Milk coagulated (B. COLI GROUP).

1. Indol produced.

a. Phenol produced.

9. *Bact. Marsiliensis.*

(B. of Marseilles swine plague, Rietsch·Jobert).

(B. of Fettchenseuche, Ebert-Schimmelbusch).

(B. of Kaninchenseptikämie, Ebert Mandry).

(Swine plague, Billings).

(Amerik Rinderseuche, Billings).

B. twice as long as broad, ⅓ smaller than *B. typhi*, polar stain, Fla peri-trichic (4-5). *Ges* typhi or coli like ; cult characters like B. coli. *M* Co+, ac+ In+, Ph+, *Lit M* reduced, red. Cults of Billings' swine plague for old cults as above, for new cults approaching hog cholera B.

Path. mice abscesses at point of inoc, do not kill old mice.

Rabbits variable, negative—slightly path. ; a general septicaemia often with the Ebert-Mandry B.

Sparrows inj. into breast muscles, death 24-35 hrs., septicaemia, pleuritis, pericarditis. (Pigeons, ⅃, hens, ferrets)+.

Hab. in blood and organs in ferret plague.

Assoc. with Marseilles swine plague, spontaneous septicaemia of rabbits.

    b. No phenol produced.

    10. *Bact. coli communis* (Escherich).

B. 0.4-0.7: 1—3, Fac an. *G cs* (1)* round-lenticular, yl-br. (2)* flat, erose-lobed, marmorated. *G st* (1)*+, beaded (2)* flat, spdg. *A sl* wh-gr. mo, gl, transl. *P,* ylsh-ylsh br. *B.* turbid+, sed+. *M* Co+.

Lit. reduced, red. Bad odor, H_2 S. *Lac.* B., gas $H \quad \dfrac{3}{Co_2} \quad \dfrac{2}{1}$

Sac. B, gas±, acetic, formic, lactic acids ; in B, NH_3, alk.

Path. variable. *Mice* 0.1-1.0 c.c. B. inper ; death 1-8 d. B. in blood and peritoneal exudate.

Guineas, 1 c.c. B inper ; death 50 hrs. ; B. in blood, peritonitis.

Rabbits. Sbc ; abscess formation.

Hab. In intestines of man and animals, fæces, water, milk ; assoc. with various diseases—peritonitis, cystitis, cholera nostras, etc.

    2. No Indol produced.

        a. Gelatin colonies of a distinctively coli type (indistinguishable from those of B. coli).

    11. *Bact. coli anidolicum* (Lembke).

Morph and in cults. like B. coli; differs in producing an amount of acid in milk intermediate between *B. coli* and *B. typhi.*

Path. mice, 0.2 c.c., B, sbc. 24 hrs, death, septicaemia.

Guineas, 0.75 c.c., B, inper ; death, 24 hrs.

Hab. faeces of the dog.

        b. *Gelatin colonies of a character intermediate between Coli and Aerogenes types (Bacteria intermediate between B. coli and B. aerogenes).*

    12. *Bact. enteritidis* (Gärtner.)

        (B. of Fleischvergiftung).

B. short, thick ; stains unequally, capsule±. *G cs* (1) br. (2) gr, transl, round, gran. *Lac.* B, gas+. P. gr wh—gr yl, gl.

Path. (mice, guineas, rabbits, pigeons, young sheep and goats)+: (dogs, cats, rats, chickens, sparrows)—.

Mice and guineas infected through the stomach, enteritis ; B. found in the organs.

* (1) after *G cs* indicates deep colonies, after *G st* growth along needle track ; (2 after *G cs* indicates surface colonies ; after *G st* growth upon surface. See Table II, VII and IX on p. 59.

Hab. isolated from flesh of a cow in meat poisoning.

    13. Bact. chologenes (Stern R.)

B. o.5: 1.3. *G cs.* wh. border erose, characters between coli and aerogenes types. *P.* wh, yl, gas+. *M* co, 1-2 d., Lac. B. gas+ ; Sac B. gas+. *Path. Mice* 0.05-0.1 c.c. B. inper, death.

Guineas, rabbits, sbc. abscess formation, and with larger doses death by inper injec.

Hab. Isolated from a case of angiocholitis with meningitis.

    14. Bact. of Ice Cream Poisoning (Vaughan-Perkins).

Ident with B. coli, but milk coagulated more quickly with strong butyric acid smell. Grows on carrot, elev, creamy, acid odor. (*B. coli* grows much less vigorously and gives no smell). *Path.* (Rabbits, cat, dogs, mice, rats)+. *Hab.* isolated from poisonous ice cream.

(Bact. *a* and *c.* Guillebeau (Ann. Micrographi IV., 225).

Like B. aerogenes but slightly motile, milk) See No. 73.

B. Milk not coagulated (Hog Cholera Group).

    1. Indol produced.

    15. Bact. icterogenes (Guarnieri).

B. and cults. like B. coli, but grows less vigorously. *Lac.* B, gas △—. *Sac.* B. gas—, M ac△.

Path. guineas, inper, septicaemia, liver degeneration.

Hab. By Guarnieri from liver and blood in acute yellow atrophy of liver. Pasquale, typhoid stools.

    16. *Bact. sinuosus* (Wright).

B. medium sized, round ends, pairs-threads. *G cs.* (2) 3d 3m.m. thin, delicate, transl, irreg, sinuous, grained brsh center→ 6m.m. with radiate foldings. *G sl.* grwh, gl, transl. *A,* limtd △. *A c.* G+. *A sl.* limtd, grwh, thin, gl. *B.* turbid, sed, no pelicle. *P* br, gr, mo, not thick, rather rough, spdg. *In.* △, (35-35°)+.

Hab. water.

    2. No indol produced.

        a. *More or less gas in lactose bouillon.*

    17. Bact. monadiformis (Messea).

      (*B coli mobilis,* Messea).

B. short rods. Fla monotrichic. Cult characters like B coli. *M* ac△. *Sac B,* gas—. *Path,* mice—*Hab.* typhoid stools.

    18. Bact. Breslaviensis (Van Ermenghem).

(Bact. der Morseeler and Breslauer Fleischvergiftung).

B 0.6-1.5 long ; slender ; 4-12 long flag. *G sc* like B coli. *B* turbid with delicate memb. *P*+, ylsh. *Sac B,* gas △.

Path. (Dogs and cats)—, mice and rabbits by feeding and injec enteritic symptoms, bacilli in organs.

Hab. isolated from poisonous beef and veal which were the cause of meat poisoning.

b. *No gas in lactose bouillon.*

19. Baet. suipestifer.

(B. of Schweinepest, Bang-Selander).

(Hog Cholera, Salmon Smith).

B. 0.6-0.7; 1.2-1.5. Fla. peritrichic. *G cs.* (1) round, br, homo. (2) round, flat, irreg. *G st.* nail, flat.headed. *P.* ylsh. *M alk. Lu M* blue. Ph—. *Path.* mice, rabbits, s bc. death 7—12d; spleen enlarged, hard, darkened, in liver necrotic spots: kidneys inflamed; B in organs. *Hab.* assoc. with hog cholera.

20. Baet. levans (Lehmann-Wolffin).

B. 0.6-1.8 Cult. characters like B coli. $Gl B$ gas $\dfrac{H}{CO_2} = \dfrac{1}{3}$

Sac. B gas—. *Gl B* lactic and acetic acids, butyric acid —.

Path. ? *Hab.* isolated from sour dough.

3. Indol production indeterminate.

21. Baet. morbificans bovis (Basenau).

B. 0.3-0.4; 1-1.2 *G cs.* (2) (a) papillate, ylsh—flat, spdg, on border with indented edges, gsh bl (b) black contour with a clear zone within, within this ylsh, mottled, gran.

G st (1) filiform (2) round, wh, thick, with undulating border. *P* mo, yl, never br. *B* turbid, with memb. *Gl B* gas _. *Suc B* gas—, *Lit M±* *Path* (mice, guineas, rabbits)+, sbc, inper, and feeding. Calves and goats, inper, feeding+: bacilli in organs and in muscles : communicated through infected meat.

Hab. isolated from the flesh of a cow with puerpural fever.

22. Baet. typhi murium (Löffler).

(B. of Mäuse typhus.)

B. like B. typhi. *G sc.* (1) small, round, gran △, yl-br, (2) typhi like. *G st* nail, flat topped· *P* whsh-grsh, △. In? Ph? *M* alk. *Path.* house mice and field mice, sbc, and feeding, death 3d. B. in organs, liver, spleen enlarged; feeding 7-14 d. *Hab* Found by Loffler as cause of an epidemic in mice.

II. No Gas generated in Glucose Bouillon (Typhoid Group).

A. Milk coagulated.

23. *Baet. equi intestinalis* (Dyas-Keith).

B. 1.0: 1.2, somewhat thicker than B. coli. $37^\circ +, 20^\circ$—, Cult. characters otherwise like B. coli. M Co. 2d. Lit M. red. *Path?* *Hab.* isolated from excrement of the horse.

B. Milk not coagulated.

1. Potato cultures whitish-grayish ; invisible.

   a. Bacteria from animal habitats. (Pathogenic. Nearly identical with B. typhi in cultural characters).

   24. Bact. meningitidis (Neumann-Schäffer).

B. and cult. characters like B. typhi. *P.* viscid. gr—wh ; *Lac.* B. gas—. *Path.* guineas + ; pyogenic reaction, sbc. *Hab.* isolated from a case of purulent meningitis.

   25. Bact. paradoxus (Kruse-Pasquale).

B. and cult. characters like *B. typhi.* *P.* spdg. invisible. *Lac B.* gas—. In +. *Path.* mice +. *Hab.* isolated from the liver from a case of dysentery in Alexandria.

   26. Bact. of Corn Stalk Disease (Billings).

B. and cults. indistinguishable from *B. typhi.* *Path.* (mice, gunieas rabbits)+ ; septicaemia ; (swine, dogs, rats, fowl)— ; (horses, goats, sheep) +. *Hab.* isolated from Billings' "Corn stalk disease ;" by Nocard from bronchopneumonia in oxen.

   b. Soil and water bacteria, not so distinctly connected with B. typi.

   * Rosolic acid solution faded.

   † Milk rendered acid.

   27. Bact. ambiguus (Wright).

B. small rounded ends. Singly, pairs-threads. *G sc.* (1) round entire, gran. brsh, darker in center (2) (a) 3-4 d, gr. transl ; elev △, irreg △, 2 m.m., sharp. (b) gran, yslh br center, thin transl margin, finely radiate. *A c G+.* *A sl,* gr., limtd, sharp. *B.* turbid, sed, pel—, *P,* thick, viscid, spdg, gr—creamy. *Lit.* *M* ac, Co only after 1 mo., and may not be then. *In* (35°—36°)+. Water.

   †† Milk not rendered acid, alkaline or unchanged in reaction.

   28. *Bact. solitarius* (Ravenel).

B. slender, straight, 3-7 x's breadth, singly, rounded ends. Ae+. *Mo+,* rotatory, non-progressive. *G cs* (1) round, grsh, gran△ ; margin notched—+ zoned, marmorated. (2) round, grsh, fllocose—filamentous ; 70 hrs. 1 m.m., gr. wh., entire, nucleus with an irreg. indistinct filamentous border and outer orange zone; 7 d. milk wh., elev, round, surface smooth, entire. *A sl.* spdg, in 4 d. porcelain wh, mo, gl, *G st* (1) filiform, (2) 4 d. elev. 1 m.m., depressed in center

beaded. *P.* thin, whsh-+ thick, pasty, color of putty. *B* turbid, no
pel. *P. R.* 7 d. decol, alk. *Lit. M.* becomes darker-+ decol. In—.
$(35°—36°)+$. Soil.
    ** Rosolic acid solution not faded.
      29. *Bact. geminus minor* (Ravenel).
B. very short rods, rounded 2-4 x's breadth, singly, Mo active. Ae +.
G cs. round, ylsh, gran, entire. (2) ylsh, gran, entire-+1od 1.5 m.m.
elev, convex; pearly wh. dense. *G st.* (1) filiform (2) 2d. a bead 1
m.m.-+larger and more spdg, with corrugated edges. *P* thin spdg-+
1od dirty wh, mo, gl. *B.* turbid, pel△. *P. R.* cherry red in 1od.
Lit M. more blue 4d.-+ 14d. slate blue, Alk+. In+. Nitrites
$(35°—36°+)$. Soil.
      30. *Bact. aquatilis sulcatus quartus* (Weichselbaum-
              Dyar).
From description (Dyar, N. Y. Acad. Sci., VIII, 95,359) indt. from
preceding.
      31. *Bact. primus Fullesi* (Fulles-Dyar).
B. 0.5·06; 0.8-1.0, singly. From description indt from 30, except *M*
cults. emit a disagreeable odor. Water.
    *** Action on rosolic acid not stated.
      32. *Ract. tracheiphilus* (Smith).
B. 0.5-0.7; 1.2-2.5. Grow poorly in gel. *A sl.* thin, smooth, mo. gl,
milk wh, limtd. *A sl* (1) lateral outgrowths 1-2 m.m. (2) surface
thin. *P* thin, smooth, wh, mo, gl. (color of potato). Does not grow
in alk media. *M—.* *B* cloudy, but not turbid. *Gl B, ac,* no gas.
Cults. very viscid 37°—. *Hab.* Diseases of melons and curcurbits.
  2. *Potato Cultures becoming yellowish-brownish.*
    a. *Grow well at the body temperature.*
     * Produce indol.*
      32. *Bact. pinnatus* (Ravenel).
B. slender, short, straight, 3-5 x's breadth, singly—short chains. Fac.
an. *G cs.* (1) round, entire, ylsh, gran. (2) punctiform, center, ylsh
br, gran; border clear, round-oval entire, do not exceed 1 m.m.-+
denser. *G st.* (1)+ (2) 2m.m., raised, porcelain wh, smooth, depressed in center; gel may become clouded. *A sl,* thin, gl, watery. *P*
thin, colorless—l dirty br., smooth, mo, gl. *B.* turbid, flakes, no pel
on surface. *P. R.* decolorized, alk. *Lit. M.* darker indigo blue. In
+; nitrites + . Soil.
      34. *Bact. viridescens non-liquefaciens* (Ravenel). See
    No. 327.

** Do not produce indol.

† Gelatin stab arborescent *i. e.* with outgrowths.

35. *Bact. geminus major* (Ravenel).

B. straight, thick, rounded, variable length, singly-short chains. Rods show deeply stained points 2-3 in. each rod. *Mot* △, Ae. *G cs.* (1) brsh, gran, entire (2) typhi-like but more gran and coarser. *G st* (1) 3 d. delicate off shoots (2) thin, spdg, with irreg borders. *A sl.* thin, transl, spdg. *P.* honey yl, mo, gl.—+ chocolate br. *B* turbid—+ clear. *P. R.* darker △ in 1od. *Lit. M.* 7d. more blue, am 5 w. In. —. Soil.

†† *Gelatin stab not distinctly arborescent.*

Milk rendered slightly acid.

36. *Bact. typhi* (Ebert-Gaffky).

B. o. 5-o. 8 : 1.3—threads. Fla. peritrichic. *G cs.* (1) round, yl br, homo, (2) roundish, undulating-lobed, marmorated, transl-iridescent. *G st.* (1) tuberculate-papillate (2) thin, erose ; *B.* turbid, less so than B. coli, memb.—*A. sl.* trans, gr. *P* yl-brsh. *M* ac △.

Path. guineas, &c. toxaemia from cults or their filtrates.

Hab. Assoc. with typhoid, stools of typhoid patients, water, etc.

37. *Bact. pseudotyphosus* (Lösener).

B. and cult. characters like 36. Diff. serum reaction.

Z. H. XXI, 238. *Hab.* Isolated by Pansini from liver abscess ; Lösener, peritoneal fluid of a hog, water, etc.

Milk rendered alkaline.

38. *Bact. faecalis alcaligenes* (Petruschky).

Indistinguishable from B. typhi except by alkaline reaction of milk (not invariably constant). Diff. serum reaction. *Hab.* faeces.

Milk reaction not stated.

39. *Bact. Friedebergensis* (Gaffky-Paak. See No. 47.

*** Indol production not stated.

40. *Bact. solanacearum* (Smith).

B. av. o.5-1.5, variable, mot △ +. Fla. several. *G cs.* (1) round, ylsh-brsh, granular, entire. (2) round, thin, wh, *G st* (1) △ (2) thin, wh limtd. *G sl.* wh, smooth, mo, gl, finger-like extensions into gel. *A sl* smooth, mo, gl, wh—+ ylsh br-br, agar stained br. *B* zoogloea in upper layer, uniformly turbid on shaking. *P* dirty wh—+ brsh-smoke black. *M co*—, slowly saponified to a ylsh transl fluid. *Lit M* becomes a deeper blue. 37° +. *Gl B.* no acid, gas—. B. alk. *Hab.* Diseases of tomato, egg plant, Irish potato.

b Do not grow at body temperature.

† Produce indol.

41. *Bact. nexibilis* (Wright):

B. medium.sized, rounded ends, pairs and long forms, chains-clumps. *G cs.* (1) round, irreg, brsh center, faintly radiate near margains +. grsh br. (2) 3d, 3m.m. thin, grsh, transl, opalescent, △ sinuous (b) brsh, gran △, margin transl, sinuous, dentate. *A sl.* thin, transl, spdg. → gsh in time. *B* turbid → faint gsh tint. *P* br, thick, viscid, spdg. *Lit. M*→ pink, acid. *In,* faint. Water.

†† Do not produce indol.

42. *Bact. aquatilis sulcatus* (No. V) (Weichselbaum). B. and in cults like B. typhi. Diff. ae in G st, little or no growth in depth. Do not reduce nitrates. Path—. Water.

43. *Bact. recuperatus* (Wright). Probably ident. with 41. Diff. ? *Lit M.* alk, color deepened. Water.

C. *Milk coagulation not stated (Bacteria of the typhoid group). Not sufficiently described to differentiate from each other and from B. typhi.*

44. *Bact. intestinus motilis* (Sternberg.)
Isolated from intestines of yellow fever cadaver.

45. *Bact. caviae fortuitus* (Sternberg).
Isolated from liver of yellow fever cadaver.

46. *Bact. cavicida Havaniensis* (Sternberg).
Isolated from intestines of yellow fever cadaver.

III. **Gas Development in Glucose Bouillon not stated (Bacteria of the Coli, Hog Cholera and Typhoid Groups not classified).**
A. *Milk not coagulated.*
1. *Pathogenic to Guinea pigs and rabbits.*
47. *Bact. Friedbergensis* (Gaffky-Paak).
B. about 1 3 smaller than B. typhi. *G. cs.* (1) round, ylsh, homo, often concentric (2) round, spdg, ylsh center paler border, marmorated, between aerogenes and coli types. *G st.* (1)+ (2) thin, spdg to walls. *A sl.* wh, gr, slimy, *In*—. *P.* whsh-gryl-rdsh. *Path.* sbc, mice, guineas, rabbits; feeding, guineas+, mice+, dogs△, cats△, rabbits△. *Hab.* From poisonous sausage. Meat poisoning.
2. *Pathogenic to pheasants, not so to guinea pigs and rabbits.*
48. *Bact. phasiani septicus* (E. Klein).
B. like B. coli but smaller; cults, coli like. *Path* pheasants, few drops B. cult, death 24 hrs, septicaemia. (fowls, pigeons, guineas, rabbits) refractory.
3. *Non Pathogenic.*
49. *Bact. Schafferi ;* (v. Freudenreich)
B. 1;2-3 threads *G cs.* (1) small round, ylsh, gran. (2) spdg, porcelain wh., irreg △. *A sl,* gr. brsh. *P.* ylsh. Hab. cheese, etc.
50. *Bact. No. 27,* (Conn.)
B. 0.8-1.3-2, no chains, 20°-35°+. *G cs.* (2) spdg. transp, elev, wrinkled edges; *G st.* (1) beaded (2) thin, transp. *A sl.* wh, elev. *P* thick, ylsh,

spdg. *M* alk, bad odor, bitter taste, pep to brsh fluid. *B* turbid, pel, sed. Milk.

B. *Milk Coagulation not stated.*
1. *Gelatin colonies of the coli type.*
   a. *Pathogenic for birds (Motile bacilli related to B. of fowl cholera).*
   * *Not pathogenic for guinea pigs.*
    51. Bact, of Canary bird septicaemia (Riek).
B. 1.2-2.5, long, polar stain. *P.* yl gr, otherwise in cults like B. cholerae gallinarum. *Path mice,* S bc, septicaemia; *Canary birds,* sooty discoloration of skin, liver necrosis, septicaemia.
    52. *Bact. diphteriæ avium* (Loir-Duclaux).
B. like fowl choera. Cult. characters not fully described. *Path* for all kinds of birds, *rabbits,* S bc, septicaemia, etc., in pharynx and larynx exudate, death 6-10d. *Hab.* cause of an epizootic among chickens, pigeons, turkeys, Canary birds, in Tunis.
   **Slightly Pathogenic for Guinea pigs.*
    53. *Bact. of Pneumonia in turkeys.* (MacFadyean).
B. like fowl cholera. *Path.* rabbits, guineas△, very weak, chickens and pigeons△, turkeys+; nasal catarrh, "rattles" in throat, pneumonia, pericarditis. B. in lungs and organs.
    54. *Bact. of Grouse Disease* (E. Klein).
B. 0.4:0.6-1.6, often coccoid. *G sc.* (1) small, round. (2) thin, spdg. irreg. *G. st.* nail, flat topped. *Path.* mice 75 per cent., guineas 50 per cent., sbc. lungs hyperaemic, hepatized△, spleen not enlarged; kidneys hyperaemic. B. in blood, lungs, liver.
In grouse, pneumonia, local hyperaemia of intestines, enlargement of liver and kidneys. B. in blood, lungs and liver.
   ***Pathogenic for Guinea pigs.*
    55. *Bact. haemorrhagic septicaemia of swans* (Florentini).
B. like fowl cholera but larger, 0.5:1.5-2-threads. *G sc.* (2b) gran△, concentric, the outer radiate-cilliate. *G st* (1) beaded (2) wh, lobed and toothed. *A sl.* round cs, coalescing, wh. *P.* cs. → coalescent, elev, yl-br, bad odor. *B.* turbid, sed wh. *Path* rabbits+, guineas+, geese+, hens+, pigeons+; coma tose condition followed by death.
In affected swans oedematous infiltrations of lungs; ecchymosis of serous mem brane, hyperaemia ^ of intestinal mucous memb; turbid degeneration of liver cells.
   b. *Not Pathogenic to birds, scarcely pathogenic to other animals.*
   * *Large bacilli.*
    56. *Bact. odematis aerobius* (E. Klien).
Bact. 0.7:1.6-2.4-24. *G cs* (1) round, brsh, (2) thin, spdg, typhi like, marmorated. *G. st.* gas—(1) beaded (2) thin, transp, dentate. *A sl.* grsh wh, smeary.
B. turbid, no pel. *P.* viscid, ylsh. *Path* cults loose quickly their virulence; fresh, first generation cults, kill *guinea* in 1 c.c. B. cult doses; bloody oedema with gas, reddening of muscles, spleen and liver enlarged. B. in oedematous fluid. *Hab.* isolated from guineas which had been inoc with faeces, earth, dust, etc.

**Small slender bacili.

57. *Bact. pneumosepticus* (E. Klein).

B. 0.3-0.4:0.8-1.6-chains, polar stain. *G cs.* (1) small, round (2) thin, iridescent. spdg, erose. *G st.*, nail, flat top, *A sl* whsh-brsh. *P.* slimy, thin, brsh. *Path, mice* 60 per cent. sbc, death ; inflam. at pt. of inoc and in lungs, spleen enlarged, hemorrhagic enteritis; *guineas*, 25 per cent. lobular pneumonia, pleuritis, etc. *Hab.* isolated from rusty sputum of pneumonia patients.

   c. *Pathogenic to insects.*

57a. *Bact. monachae* (Tubeuf).

B. 1:0.5 singly, in 2's or short chains. *Ae. Gel. cs.* (2) (a) transp. opalescent, mother of pearl lustre (b) central portion ochre yl, sometimes zoned. Edge erose, lobed. *B* turbid. *P* mo, gr. *Path* infection expts. positive. *Hab.* in body fluids of diseased " Nun moth'' larvae. (*Liparis monacha*).

2. *Gelatin colonies of the aerogenes type.*

   a. *Pathogenic to the smaller animals.*

     *Rabbits. *General infection.*

     58. *Bact. cuniculi septicus* (Lucet).

B. 1-3. *G cs* (2) smooth, very convex, slimy. B. 39°-40°, stringy masses. P. no growth. *Path. Rabbits*, 0.25 c.c, sbc, inper, 24'h. death ; spleen enlarged, serous membr. inflamed, local oedema, B in all organs. *Guineas.* sbc, abscesses ; inper, death ; chickens and pigeons refractory. *Hab.* assoc. with a spontaneous epizootic of rabbits.

     59. *Bact. venenosus* (Vaughan)

     (Bact. venenosus brevis, Vaughan).

B. 2-4 x's, breadth, rounded ends, *G cs* aerogenes like. *A sl* thin. wh. *P.* mo, gl, lbr. *Pariotti's sol* ⊦. *Uffelman's Gel.*÷-∆. *Path* rats+, mice+, guineas �, rabbits +. Water.

     **Pyogenic for smaller animals, etc.

     60. *Bact. glischrogenum* (Malerba).

Characters of B. aerogenes. Milk and urine rendered slimy. *Path* pyogenic to smaller animals. Causes nephritis in dogs. *Hab.* isolated from urine.

   b. *Non-pathogenic.*

     *Do not grow at 37°C. *Water Bacteria.*

     †Growth on agar, smooth, not characterized.

     61. *Bact. albus* (Eisenberg).

B. short. *G cs.* small, aerogenes like. *P* rugose, ylsh wh, limtd. Water.

     62. *Bact. aquatillis solidus* (Lustig).

B. 3x's width-filaments. *G cs* (2) aerogenes like, gran. *P.* grwh-ylsh. Re duces nitrates. Water.

     ††Growth in agar, branched (*Wurzelartig*).

     63. *Bact. stolonatus* (Adametz).

B. 2-3 x's breadth. *G cs* (2) semi-sph, whsh-brsh. *A sl* large cs, wurzelartig. *P.* dirty wh.

     **Grow well at 37° C.

     64. *Bact. denitrificans agilis* (Ampola-Garino).

B. 0.1-0.3:1.—1.5. *G cs.* small raised, gran, ylsh. *G st.* (1)_∆. *A sl* dirty wh. limtd. *P* _∆, ylsh-dirty br. Gas in Gel. and B containing nitrates; gas N. 13, Co_2. 2 *Hab.* Isolated from manure.

3. *Colonies not characterized as above.*

    65. *Bact. denitrificans I* (Stutzer-Burri).

    B. 0.75:1.5—2.5. G cs (1) small, round (2), weak growth. *G st.* (2) thin, grsh. *P.* br-rdsh. Reduces nitrates. Horse manure.

    66. *Bact. denitrificans, II.* (Stutzer-Burri).

    B. 0.75:2-4. *G cs.* (1) small, eliptical lacerate, radiately-reticulately rugose. *P..* rugose, slimy, rdsh. *B.* membr. *Hab.* isolated from straw.

4. *Colonieg burr-like.*

    66. Bact. invisibilis (Vaughan).

    B. large, rounded ends, 2-5 x's breadth, 20°.37°+. *G cs* pale yl, burr-like with irreg outlines. *G st* (1) + (2) △. *A st.* thick, wh. limited. *P.* invisible. *Pariettis* sol +. *Uffelmann's Gel.* +. Water.

    67. *Bact. venenosus invisiblis* (Vaughan). Not clearly diff from above.

CLASS III. GELATIN NOT LIQUEFIED. NON-MOTILE.

I. Obligate Aerobic (Acetic Ferment Group).

A. Form long chains of more than four elements, also long Involution Forms.

    1. On sterile beer or beer wort a membrane over entire surface.

    a. Stained yellow with Iodine solution.

    68. *Bact. aceticus* (Hansen).

    (Mycoderma aceti).

B. short rods, in chains, involution forms. *Wort G cs* aerogenes type-rosette like-radiate. *Sterile beer* 34°, 24 hrs., memb moist, slimy, smooth, veiny. B. in chains. *P.* no growth. op. 30°-34°; Max. 37°; min 10°. *Hab.* wine and beer.

    b. Stained blue with iodine.

    69. *Bact. Pasteurianus* (Hansen).

B. as before. *G cs* entire, without any rosette form, but with brain-like corrugations of surface.

Sterile Beer 24° 24 hrs. memb dry, minutely corrugated B. longer and broader than preceding (?). *Hab.* Beer. beer wort, seldom in wine.

    2. On sterile or beer wort, an imperfect membrane as islands.

    70. *Bact. aceticus* (Zoidler).

B. morph. like *B. aceticus* (Hansen) *Wort G cs* (1) round, center, gran △, brsh yl with smooth transp border (2) ylsh drops, pear-shaped. *P.* scarcely observable. • *G st* (1)-(2) +. *Hop beer wort* turbid yl br; memb of islands. *Hab.* Beer wort.

B. *Do not form chains of more than four elements.*

    71. *Bact. aceticus Petersii.*

B. 0.8:1.6, in chains 2-4 elements at the most. *G cs.* round, homo, flat, spdg. In yeast water with 5 per cent. alcohol grows well, turb with memb. *Hab.* isolated from sour dough.

II. Aerobic and Facultative Anaerobic.

1. Gas generated in glucose bouillon.

   1. Gas generated in lactose bouillon. (AEROGENES GROUP.)

      a. *Milk rendered viscous, slimy.*

      72. *Bact. viscosus cerevisiae* (van Laer).

B. o.8:1.6-2 4 *G cs.* entire, erose, brsh. *P.* wh, watery, doughy cs, which smell like foul fish. *Beer wort,* viscous, Co_2 evolved. *M* and *Gl B* slimy, gas. *Hab.* isolated from beer, yeast, bread, cause of a viscous fermentation.

   b. Milk coagulated but not rendered viscous-slimy.

      *Bouillon rendered turbid.

      73. *Bact. aerogenes*

         B. lactis aerogenes, Escherich).

B. o.5-1.o:1-2 *G cs.* (1) round, gran, gr. br (2) porcelain wh., round (b) ylsh, gran. darker in center. *G st.* nail, round head. *B.* turbid, memb$\triangle$. *A sl.* opaque $\searrow$, porcelain wh. *P* mo, ylsh wh, gas, chessy ordor. *M Co -* , ac. In—.

Path. variable. For smaller animals usually pathogenic only in large doses (toxaemia).

Var. Capsule B. of Nicolaier. *Mice* sbc—. : *rabbits* sbc, local supuration; *guineas*, inper, fibro-purulent peritonitis, some invasion of B. into blood, and death within 24 hrs. B. in peritoneal exudate, with a capsule.

Hab., faeces, milk, cheese, water, air, associated with cystitis.

74. Variety. B. *acidi lactici* (Grotenfeld).

Indistinguishable from the descriptions from the preceding. *Hab.* faeces, water, milk.

75. B. a. and b. Guillebeau.

Indistinguishable from 73, expect B. show a slight motility (?). Milk· ** Bouillon not rendered turbid.

76. *Bact. pallescens* (Henrici). From descriptions indistinguishable from 73 except B. is not rendered turbid. *Hab.* isolated from cheese.

Note.—Under this head is placed (771. *Bact Coli immobilis*, from faeces; indistin guishable from B Coli, except as to motility. See No. 10.

2. No gas generated in lactose bouillon (FRIEDLANDER BACILLUS GROUP.)

   a. Milk coagulated.

      * Rosolic acid solution decolorized.

78. Bact. sordidus. (Dyar.)

B. o.6-1.o: 1-1.5, singly-pairs, rarely short chains, rarely short chains, rarely a capsule. *M* co, nitrites △. *A sl* viscous growth. *Lac. lit* reddened, later blue. *Hab.* Described by Dyar as *Micrococcus sordidus* from Kral's laboratory.

    ** Rosolic acid solution not decolorized.

    79. *Bact. acidiformans* (Sternberg-Dyar).

B. o.6: o.8-1.o, short B. resembling cocci. Nitrites △. *Lac. Lit.* reddened → 20-90 d gradually blue.

    b. *Milk not coagulated.*

    * Bacilli short with a capsule in animal body. Does not grow out into very long threads.

    † Colonies watery, transparent, approaching the coli type.

    80. Bact. ozaena (Abel).

    (B. mucosus ozaena, Abel).

    (B. capsulatus mucosus, Fasching).

    (B. rhinitis atrophicans, Paulsen.)

G st. (2) more spreading than B. aerogenes. *P.* 20°, watery, thin ; 37°+, ylsh, opaque. Does not grown in acid gel. like No. 73.

Path. sbc., death, 1-4d.; spleen enlarged B. in blood ; guineas, sbc—; inper, death; heptization of lung, peritonitis, exudate stringy. *Hab.* isolated from nasal secretions.

    81. *Bact. rhinosclermatis* (Paltauf).

B. and cults. indistinguishable from preceding. *Path.* generally slight for smaller animals, like *B. pneumoniae. Hab.* assoc. with *rhinoscleroma.*

    †† Colonies white, nearly or quite opaque, aerogenes type.

    82. *Bact. pneumoniæ* (Weichselbaum).

(Pneumoncoccus of Friedlander, Friedlander's B, Capsule Bacillus.)
Cults. like B. aerogenes ; Gel. stained br. Gas △, $\frac{H}{CO_2} = \frac{3.}{2.}$

Path. mice, sbc, better inper ; septicaemia, lung hepatized. *Rabbits* immune. (The large number of probable synonyms of this species are given in the index.)

    B. Very little or no gas generated in glucose bouillon.

    1. Milk coagulated.

        a. *Gelatin colonies of the aerogenes type. Bacteria closely related to B. aerogenes No.* 73.

        83. *Bact. limbatum acidi lactici* (Marpmann).

B. short, thick, with a capsule. *M. serum*, *G cs*, 24 hrs. milky, punctiform, gl., edge sharp. G st. (1)△, (2) flat, wh, *Lit M* 24 hrs. Co, a
slight reddening. 20°-37°. Milk.

    84. *Bact. ureae* (Jaksch–Dyar).

B. 0.7·1.0: 1-2. Singly and in short chains. *A sl* thin, wh-transl, spdg.
P R+. In+. *Hab.* described by Dyar, a specimen from Kral'a
laboratory.

    b. *Gelatin colonies of the coli type.* (Fowl Cholera Group.)

    *Produce indol.

    85. *Bact. cholerae gallinarum.*

    B. der Huhnercholera.

    B. der Kaninchenseptikämie (Koch-Gaffky).

    B. cuniculicida, Flügge.

B 0.4-0.6: 1.0, polar stain, *G cs* (1)—(2) like B. coli. *A sl* aggregations
of delicate cs. *P.* at 20 ·37° waxy, transl, gr wh, flat. *B.* slightly
turbid. *M Co*+, ac. *Lit M.* reduced. In+. Ph+.

Path. sbc. injec of small doses, septicaemia in chickens, pigeons, geese,
ducks, etc., rabbits, mice.

Hab. assoc. with chicken cholera.

    86. *Bact. gallinarum* (E. Klein.)

B. 2 x s length of fowl cholera. Cult. characters ident. with 85. *P* 37 ,
no growth, later a brsh growth. Diff. from preceding mainly by its
weaker *Path*; only chickens affected sbc, and by feeding; loose stools,
death 7-9 d; B. in blood.

Hab. isolated from cases of enteritis in fowl.

    ** Do not produce indol.

    87. *Bact. coli anaerogenes* (Lumbke).

B. 1.0: 2.0. Morph. and cults. like B. coli. *Lac. B*, gas—, acid+.
Acid production between B. typhi and coli. *Path.* sbc. injec, septicaemia, mice, guineas, rabbits, B. in blood. *Hab.* isolated from the
faeces of a dog.

    2. Milk not coagulated.

        a. Lactose litmus milk rendered acid, or reddened.

        * Small bacilli about 0.5 u wide.

        † Produce indol.

        88. *Bact. suisepticus.*

        (Der Deutschen Schweineseuche, Schultz).

        (Swine plague, Salmon-Smith).

B. morph. and in cults. like 85. *M.* ac+, *Lit.* not reduced. *Path*, mice,
rabbits, small birds, sbc. injec, death septicaemia, 24 hrs. Chickens

more refractory to small doses than 85, large doses fatal. Swine, sbc., injec, oedema at point of injec. and septicaemia. *Hab.* assoc. with above disease.

89. *Bact. bovisepticus.*

(B. Wildseuche, Hueppe.)

(B. Rinderseuche, Oreste-Armani.)

(B. Buffelseuche, Oreste-Armani.)

B. and cults. indst, from preceding. See Canvena C.ix, 561. Ph (—) for Buffelseuche ; In +, Ph + for the other varieties. *Path.* Differs from preceding only by its pathological characters.

90. *Bact. felis septicus* (Fioca). See No. 114.

†† Do not produce indol.

91. *Bact. pestis bubonicae* (Yersin-Kitasato).

B. short ovals, 1.0 to long-rods 4·5 x's breadth, also chains of short elements, often with a capsule. Gm ?, polar stain. *G cs* (1) round, whylsh wh (2), flat, does not grow larger than a pin's head, gran, border. *G sl* dry wh-ylsh, gran △. *G st* (1) ⊥ (2) flat. *A sl.* confluent cs, stringy, *P.* △, wh-gr. *B.* floc. adhering to walls or turbid with much sed. *M* △. Grows on slightly acid medium. *Lit M.* reddened in 24 hrs.

Path. (mice, rats, guineas, rabbits) +. Oedema at point of inoc., swelling of lymph spaces, congestion of inner organs. B. in organs and blood, with death in a few days. *Hab.* isolated from superating glands in bubonic plague.

92. *Bact. sporadic pneumonia in cattle* (Smith).

B. like B. of swine plague except presence of a capsule, 0.5-0.6:10. Cults, become viscid with age. Growth on gel.\ or invisible. *A cs* (2) round, transl , grsh—+ 4.5 m. m. *A sl*, grsh, gl, fleshy , condensation aq+- dense, viscid. *P.* not manifest. *M±.* *Ph.+*, In—?, generally absent. *Gl B*, *sac B*, ac, no gas, *Lac B±.* *B* \ deposit, viscid when old ; no gas. *Path.* like swine plague.

†††Indole production not stated.

93. *Bact. sorracenicolus* (Dyar).

B. small, 0.5-1.0, singly, pairs and short chains. *A sl* whsh, spdg +. ylsh, . *Lac. Lit.* red—+ 4)-90 d, blsh. Nitrites—. *Hab.* from fresh leaf of the pitcher plant.

**Large bacilli, 1.0 u. wide.

94. *Bact. vacuolatus* (Dyar).

B. 1.0:1·5 u, rounded ends, vacuolated, singly-short chains. *Mo—*, occasionally spasmodically so. *A sl* gl wh, transl. *Lac. Lit* red—-+ 6od, blsh. Nitrites—. *Hab.* From a trap of the carnivorous water plant *Utricularia vulgaris.*

b. *Lactose Litmus Milk, blue, reaction, amphoteric-alkaline.*

   * Gelatin colonies smooth, or not characterized as below.

      † Produce indol.

        95. Bact. sanguinarium (Smith).

B 1-1.3 : 2-1.8, ends tapering-rounded, pairs—clumps, polar stain. *G cs* (1) 0.25 gran ⌐, homo, (2) gran, spdg , no markings. *G st.* growth more abundant along line of inoc than on surface. *A sl* 37 grsh, gl. *P.* grsh yl, delicate. *B.* turbid, sed. *Gl B.* acid. *M.* 4 w. ⁖ —+ alk. *Lac. B-Suc. B.* no acid, *Ph* —. *Path.* chickens, pigeons, rabbits, mice. *Hab.* assoc. with infectious leukaemia in fowls.

        96. Bact. decolorans major (Dyar).

B. 0.7-0.6 : 1-2 singly-chains. Nitrites—, *P R* decolorized in 16-28d. Cult. characters not fully described. *Hab.* isolated from the air.

        96 a. Bact. decolorans minor, Dyar.

Differs from 96 in reducing nitrates slowly ; B. also smaller, 0.6:0.6-1 o.

      †† Do not produce indol, or reaction doubtfully faint.

        97. Bact. lactis innocuus (Wilde).

B and in cults. similar to B. aerogenes. *G cs.* porcelain wh, round-irreg. with characters approaching B. coli. *P* brsh. *Hab.* milk.

        98. Bact. tiogensis (Wright).

B. medium, plump, singly-pairs-short chains-threads. *G cs* (2) ² d. 2 m. m., rounded, milk wh., elev, gl (2b) dark, opaque, gsh schimmer -+ thinner, brsh, gran towards margins where they are nearly clear. *Ac G ·*, *A sl* gr, gl, limtd ⌐. *B.* turbid. *P* dirty, brsh, gr. spdg. *Lit M.* decol, am. *In ?* 37°—. Water.

      †††Indol production not stated.

        99. Bact. secundus Fulessii (Dyar).

B. 0.7-1.0:1.2, singly. Nitrites -. *P.R.* not faded. *A sl*, wh. thin. *Hab.* isolated from the air.

    **Gelatin colonies crimpled-scolloped-petaloid.

      †Colonies radially crimpled.

        100. Bact. refractans (Wright).

B. short, thick, medium sized, pairs-clumps. *G cs* (2) ² d. 1 m.m., round, wh, elev ⌐, (2b) brsh, segmented—, radially crimpled, scolloped outlines. *G sl* narrow, wh, wrinkled. *A sl*, narrow, thin, transl, c s. *P. R.* ⫽or lighter ? *B.* turbid , pel , sed. *P* . grsh-brsh gr of small cs. *Lit M.* no change. In—, 37° — .*Hab.* Water.

††Colonies petaloid.

101. Bact. rodonatus (Ravenel).

B. short ovals, rounded, like B. prodigiosus. *G cs* (1) (b) ylsh br, with irreg edges, rosette structure (2) 60 hrs. 1 m.m. grow slowly with a rosette structure as in (1) becoming distinctly petaloid on the edges; rdsh br in center, ylsh gr on edge. *A sl* wh, transl, thin, limtd. *G st* (1)+(2) thin, irreg, leafy. 4-5 m.m. *P.* mo, gl, ylsh→ brsh. *B.* turbid, pel thin, friable. *P. R.* decolorized. Lit. M. pure blue 3-4 d + decolorized 10 d. In—, 37°+. *Hab.* Water.

3. Milk coagulation not stated.

   a. *Peptone Rosolic Acid decolorized, Cultures viscous.*

102. Bact. Zürnianum (List-Dyar).

B. 0.5:0.6-1.0, singly-short chains. Cults. on solid media transl, wh., very viscous drawing out into long threads. In—. Nitrites △. *Hab.* isolated from the air.

   b. *Peptone Rosolic acid not decolorized. Cultures not decidedly viscous.*

    * Gelatin colonies with nipple-like projecting center, reticulately marked.

103. B. Martizezi (Sternberg-Dyar).

B. 0.5:1-1.5, singly-short chains. *Nitrites* ., 28d. *A sl* transl, wh. *P. R.* deepened in color.

Hab. liver yellow fever cadaver, Sternberg; air, Dyar.

    ** Gelatin colonies, thin translucent.

104. Bact. inutilis (Dyar).

B 1.0:1-2, singly-pairs, 37°+. Characters all negative ; Cults. on solid media very thin, transl. *Hab.* Air.

   4. Milk coagulation not stated (Bacteria closely related or identical in cultures in either fowl cholera or swine plague bacteria.)

   a. *Strongly pathogenic for rabbits.*

    * *Distinctly pathogenic for guinea pigs.*

    105. *Bact. cuniculicida thermophilus* (Lucet).

Path. rabbits—, guineas+, sbc and feeding, death 1-3 d, septicaemia, spleen and liver enlarged, serous membranes inflamed. B. in organs. *Chickens* immune.

Hab. assoc. with an epidemic of rabbits and guinea pigs.

    106. *Bact. of Buffalo Plague* (Ratz).

Path. (cattle, horses, swine, guineas, mice, pigeons)+ *rabbits* sbc, temp. elevation, 40°—41.3° within 8-11 hrs. ; animal dies in convulsions, bloody serum in body cavities, mesentery and serous membranes injected, organs engorged.

In air passage bloody mucus; spleen enlarged, etc, *Guineas,* sbc, abscess at-
point of inoc, death in 6-14 d. Autopsy as above; some guineas refractory.

**But slightly pathogenic for guinea pigs.*

107. Ba t. *cuniculicida immobilis* (Smith).

Path. mice, guineas, pigeons, slightly virulent, death only with large doses. For
rabbits, death, inflammation of serous membranes. *Hab.* as the cause of a
spontaneous rabbit plague.

b. *Pathogenic for rabbits only by intra-peritoneal injections.*

108. *Bact. gracilis cadaveris* (Sternberg).

B. 1 2 chains, *G st* (1) beaded below, branching outgrowths above (2) thick,
wh. *P.* creamy. *B.* turbid, bad odor. *Path* for rabbits only by inper injections. *Hab.* isolated from human liver, cadaver.

C. *Gas production in glucose bouillon not determined.*

1. *Colonies on gelatin of the coli type.*

a. *Pathogenic bacteria.*

* *Bacteria closely related to the B. of fowl cholera,* (**Septicaemia
Haemorrhagicae group**).

† *Decidedly pathogenic to rabbits, with general septicaemic infection.
Associated with specific diseases of pigeons.*

109. *Bact. diphtheriae columbarum* (Loffler).

B. somewhat longer and more slender than B of fowl cholera. *G cs.* typhi
like. *A sl* gr, transl. *B.* turbid. *P.* wh-grsh, ln—. *Path* (mice, rat bits)
+.; death with necrotic spots in liver, containing bacilli; spleen enlarged.
Pigeons infected through wounds of mouth, diphtheritic deposits containing
B. death 1-3 w.

B. in organs after death. *Hab.* assoc. with diphtheria in pigeons. Only local
necrosis in guineas.

110. *Bact. cholerae columbarum* (Leclainche).

B. somewhat larger than B. of fowl cholera, cults, like the latter. *B* not turbid, but a floc. sediment. *P.* grsh yl. *Path.* rabbits, sbc. death 8d;
guineas · 8d. By feeding wild pigeons cults. death 3 6d. with symptoms of
chicken cholera.

Diff. from 85 and 109 by *path.* in guineas; growth in *B,* etc. *Hab.* assoc.
with a disease of wild pigeons.

Associated with specific disease of rabbits.

111. *Bact. diptheriae cuniculi* (Ribbert).

(B. of intestinal diphtheria of rabbits).

B. 1-1.4:3-4. *G cs* coli like. *P.* flat, whsh, spdg... ln—. *Path.* rabbits,
sbc, inper, death 3-14 d, in liver and spleen necrotic spots containing B. Infection *per os* a diptheritic inflammation of intestines, &c. *Hab.* associated
with above disease.

112. *Bact. cuniculi pneumonicus* (Beck).

(B. of lung plague of rabbits).

B. like fowl cholera, polar stain. *G cs* small entire, gran, clear + brsh *G st*
(1) . *P.* 20° no growth. *A sl* porcelain wh-brsh. *Path.* rabbits moc. into
lung, cough, fever, rapid respiration; death 3-5 d. pneumonia, pleuritis
much exudate containing B; sbc. injec. spreading necrosis, death without

general infection. *Guineas* as above. *Diff.* from above by growth on potato and pathogensis. *Hab.* Assoc. with a lung plague of rabbits.

Septic bacteria of mixed origin.

Produce general septicaemia in guineas, strongly pathogenic.

113. *Bact. dubius pneumonicus* (Bunzl-Federn).

B. short rods, polar stain—, longer and more slender on agar. *G cs*, spdg △. *A sl*, transp cs. *P.* no growth. *Path.* (rabbits, guineas, mice, pigeons) +, sbc, inper, death 1-4 d, septicaemia, local œdema-necrosis. *Hab.* rusty sputum of a pneumonia patient.

114. *Bact. felis septicus* (Fiocca).

B. short rods. Cults. like B. of fowl cholera. *M co*—, *P.* thin, invisible. *Path.* (mice, guineas, rabbits) · . Septicaemia.

115. *Bact. septicus agrigenus* (Flügge).

B. and in cults. similar to B. of fowl choelera. *Path* (mice guineas, rabbits) +. *Hab.* isolated from earth.

(B. septicus hominis, B. of okada, B. canalis parvus, not easily diff. from 115.)

Slightly or negatively pathogenic to guinea pigs.

116. *Bact. haemorrhagicus sepricus* (Babes).

B. short rods, with a capsule. Grows sparingly in Gel. *A sl* small, transp drops → whsh yl. *P.* whsh drops. *B.* turbid. *Path. mice*, death few days ; septiaemia haemorrhagicae. *Rabbits*, death 3-8d. congestion of lungs, liver, spleen. For guineas and dogs, but slightly path. *Hab.* isolated from a case of septicaemia in man.

117. *Bact. coprogenes parvus* (Bienstock).

B. and in cults, like B. of fowl cholera. *Path.* mice, sbc, 36 h. death, oedema, few B. in blood ; *rabbits* inoc. in ear, erysipelas, diarrhoea, death in ± 8 d. *Hab.* isolated from faeces.

††*Less strongly pathogenic to rabbits. Associated with hemorrhagic infection of men.*

118. *Bact. haemorrhagicus* (Kolb).

B. o.8:1.2-threads, capsule±. *G cs* flat, erose. *P mo*, thin, *Path, mice*, death 2-3 d, septicaemia ; guineas infected only by large doses ; *rabbits* often die by inper injections of o.5-1. c.c. *Hab.* isolated from the corpses of persons dead of septicaemia.

119. *Bact. haemorrhagicus velenosus* (Tizzoni-Giovannini).

B. o.2-0.4:0.7-1.3 *G cs* flat, irreg, flocose borders. *P* invisible. *Path.* (dogs, rabbits, guineas+) causes only local oedema, with fever, hemorrhagic nephritis, vomiting, blood diarrhoea, spleen normal, necrosis of liver and epithelium of kidney. *Hab.* isolated from a case of purpura haemorrhagica.

120. *Bact. haemorrhagicus nephritidis* (Vassale).

B. similar to B. of fowl cholera, but less pathogenic for rabbits, strongly so for guineas, inper inoculations cause hemorahagic nephritis. *Hab.* isolated from a case of hemorrhagic nephritis.

†††*Not pathogenic to rabbits.*

121. *Bact. aphthosus* (Siegel).

B. short, o.5.0.7—threads, polar stains. *G cs.* small, entire, bl wh-ylsh, *G st.* beaded. *Path.* (rabbits, guineas, mice, dogs, cats)—, local infection through the mouth to young pigs and calves.

Hab. isolated from liver and kidneys of man and cattle with "Maul and Klauen-seuche."

122. *Bact. dysenteriæ vitulorum* (Jensen).

B. somewhat larger than fowl cholera, polar stain, cults. like B. coli. *P.* slimy, brsh.

Path. feeding of cults., 5 c.c. B. to very young calves, gives a fatal diarrhœa. B. in intestines and organs. *Hab.* assoc. with dysentery of calves.

** *Bacteria slender, minute, influenza like.*

123. *Bact. salivae minutissimus* (Wilde).

B. small like influenza B. *Gm—*, *G st.* nail, flat top. P. brsh. *Hab.* isolated from secretions of mouth.

b. *Non-Pathogenic Bacteria* (Milk Bacteria).

* *Milk not coagulated.*

† *Milk rendered decidedly acid.*

124. *Bact. No. 56*, Conn.

B. o.8-1.2, pairs-chains (20°-35°) , *G cs* large wh, thin, transl, irreg-lobate, irreg. surface. *G st* (1) + (2) thin, irreg. border. *A sl* thin, wh, spdg. *B.* turbid △, sed. wh. *Hab.* milk.

†† *Reaction of milk unchanged.*

125. *Bact. No. 55*, Conn.

B. o.8: 1.0, 35°△. *G cs*, wh, spdg, gran, entire. *G st* (2) thin, transl, spdg. *A sl* wh. *P.* elev, ylsh. spdg. B. clear, flaky sed.

2. *Colonies on gelatin of the aerogenes type, non-pathogenic.* (*Mostly milk bacteria probably of the aerogenes group. See Nos. 72-75.*)

a. *Bacteria grow in a greater or less degree along the track of needle in gelatin stabs.*

* *Milk coagulated.*

† *Milk coagulated at room temperetures.*

126. Bact. No. 19 (Adametz).

B. B. o.8: 2-3—threads, *Cs*, compact 4 6 d. scarcely visible points, opaque—d br, gran, not spdg. *G st* (1) △. *M ac. Hab.* Milk.

126a. *Bact. No. 53*, Conn.

B. short (20°—35°)+ ; *G cs* raised bead, 1 m.m. *G st.* (1) · (2), spdg△, pearly wh, waxy, gas in gelatin. *Asl* elev, wh, *P* wh-brsh. *M.* ac, pep—. *B.* turbid△, sed ⌐. *Diff.* 126 and 126a may be identical; 126a grows well along needle track and 126 poorly, as judged by descriptions. *Hab.* Milk.

†† *Milk not coagulated at room temperatures, only at ~ 35° C.*

127. *Bact. No. 16*, Conn.

B. ovals—coccoid forms, 1.0 (20°—35°)+. *G cs* wh-bead + spdg, thin, sometimes a raised central necleus, *G st.* nail shaped, gas in gel, frequently. Asl thin, wh, gl; spdg-elev. M co ⌐ at 37°, gas ₁, ac , pep. *Hab.* Milk.

** *Milk rendered slimy.*

128. *Bact. lactis pituitosi* (Löffler).

B. rather thick rods, breaking into coccoid segments. *C cs* round, entire erose. o.2-o.5, m. m. (b) brsh, radiately striped. *P* gr wh, rather dry. *A sl* dirty wh, cs. *M ac*, slimy. *Hab.* milk.

*** *Milk not coagulated.*

129. *Bact. No. 41* (Conn)

B. 1.1:6, in 2's, op. 20°—37°ͺͺ. *G cs* bead like, 1 m.m. *G sl* (1)△ (2) mo, elev, areogenes type. *A sl* wh, gl. P. elev, wh-ylsh. *M* ac ͺ, pep△. *B* turbid, pel+. *Hab.* Milk.

3. *Colonies on gelatin not characterized as above. Milk not coagulated. Rendered alkaline.*

    a. *Milk after 6—8 weeks becomes pasty, brownish.*

      130. *Bact. No. 26* (Conn).

B. 1:4, variable-chains, polar stain. *G cs* rosette, dark center, gran, rays-rosette lobing. *G sl* (1)ͺ. (2) thin, transp, rough, dotted with opaque spots. *A sl* very wb, elev. P gr wh. spdg ͺ. *M* co—, alk, pep+, →pasty br. *B.* clear→, turbid, pel irridescent. *Hab.* Milk.

    b. *Milk not becoming pasty or brownish.*

      131. *Bact. No. 22* (Conn).

B. short oval rods, 1:1.2.2, 35°△. *G cs.* elev, spdg, central nucleus, transp edge, radial lines. *G sl* (1) + (2) elev, wh opaque, sinks slightly in gel. *A sl* spdg △, dry, opaque. transp. edges. *P* wh, elev. limitd → ylsh. *M co*—, alk → transl △. *B.* turbid, pel—, sed. *Hab.* milk.

CLASS IV. GELATIN NOT LIQUEFIED, STAINED BY GRAM'S
METHOD. MOTILE.

I. Colonies on gelatin. Flat spreading. Coli type.

A. Gelatin surface colonies distinctly coli like. Rather thick, yellowish-brownish by transmitted light, or granular.

132. Bact. muripestifer (Laser).

(B. der Mäuseseuche, Laser.)

B. short rods, polar stain; fla peritrichic. *G cs* (1) round, brsh, (2) spdg, coli like. *G st*, nail, flat top. B. turbid, memb. △. *P* brsh. *Lit. M*, ac +. *Sugar B.*, gas+. *Path, mice* and *field mice*, sbc, 2d death; by feeding 3-10d; B. in all organs; *rabbits, guineas, pigeons,* sbc and inper, death. *Diff.* very similar to *B. typhi murium*, but distinguished by the Gram reaction. *Hab.* assoc. with a plague of field mice.

133. Bact. coli colorabilis (Naumyn).

(B. cuniculicida Havaniensis, Sternberg.)

B. short thick like B areogenes, often in 2's and short filaments. *P.* gr-ylsh br.

Path, mice, sbc, death, septicaemia, *guineas* slightly affected. *Rabbits —?* *Hab.* isolated from contents of gall bladder, faeces, yellow fever cadaver.

B. Gelatin surface colonies thin, translucent.

134. Bact. exanthematicus (Babes-Oprescu).

B. 0.3-0.5, thick, often very short and in 8-shaped forms. *G cs* (1) round, ylsh-br. (2) spdg, .rreg, transl, whsh. *Asl*, gl, gr, transl. *P*. gr-brsh, transl. *B* turbid, sed, memb. *Path* (mice, guineas, rabbits, pigeons)+, death 2-4 d, local inflam, enlargement of spleen, brownish color of organs, B. present.

Hab. isolated from a case of hemorrhagic infection in man.

135. Bact. accidentalis tetani (Belfanti-Pescarolo).

B. small, short, polar stain±. *G st* (1) beaded (2) thin, iridescent. *P*. ylsh, gl.

Path, (mice, guineas, rabbits)+, (pigeons, chickens, geese)—. Death in a few d., B. in blood ; spleen swollen ; often paralysis with convulsions. *Hab*. isolated from the wound pus of a person dead of tetanus.

II. Colonies on Gelatin of the aerogenes type.

136. Bact. endocarditidis griseus (Weichselbaum).

B. typhi like in size, diphtheria-like in form. *G cs*. aerogenes like, like Friedlander's B, but of a grayer color. *P* +, dry, g-yl br. *Path, mice* and *rabbits*, sbc. local inflam-supuration. *Hab*. isolated from a case of endocarditis.

GLASS V. GELATIN NOT LIQUEFIED. STAINED BY GRAM'S METHOD. NON-MOTILE.

I. Gas produced in Glucose Bouillon. (Bact. acidi-lactici Group.)

A. Milk coagulated.

137. Bact. acidi lactici (Hueppe).

B. 0.5-0.6: 1-2, often in 2's, optimum 37 ' *Gcs* (1) coli like. *Gst* nail, flat top. *A sl*, wh. yl. *P*. yl. br. *M co* , ac · , CO_2, alcohol. *Lac. B*.. gas · . *Path*—. Milk.

B. Milk not coagulated.

138. Bact. endometritidis (Kaufmann).

B. medium sized, length variable, with a capsule. *Gcs* coli like. *P*. ylsh. *In*—. *Lac. B*., gas ; *Suc B* gas—, *Path*? *Hab*. isolated from liver abscess.

II. No Gas produced in Glucose Bouillon.

A. Milk coagulated.

139. Bact. lacticus (Gunther-Thierfelder).

B. 0.5 0.6: 1 u, in 2's or short chains. *Gcs* small prominent. *P*. scant development. *Gl B*. acid . *Lac*—, *B*, gas ac. *M co* aromatic, alcohol and fatty acids. *Hab*. Milk.

B. Milk coagulated (28 d.) only after boiling.

140. Bact. sputigenes crassus (Kreibohm).

B. short thick-coccoid ; in body with a capsule. *G cs* aerogenes like, gr wh, gran, large. *G st*, nail, round head. *P*. gr wh, mo. *M Co* ı in

28d, only after boiling (Dyar). *Path, mice*, sbs, 2d death, septicaemia. *Rabbits* as before with inven injections. *Hab.* isolated from sputum.

III. Gas Production in Glucose Bouillon not Stated.

A. *Oligate Aerobic. Grow very slowly.*

    141. *Bact. urea* (Luebe).

    B. 1:1.5-2.6, grow very slowly. *G cs* (2) flat, spdg, irreg. Ferments urine. *Hab* isoluted from urine.

B. *Not Strictly Aerobic as above.*

    1. *Grows best on blood serum at 37° C.*

    142. *Bact. acnes contagiosae* (Grawitz-Dieckerhoff).

    B. very small, long-oval, rods, short chains, singly ; stains with difficulty. *G st* 17° and above, beaded along needle track. *P.* scarcely any growth.

    Path. horses, by rubbing cults. into skin ; characteristic pustules. *Rabbits*, do, sbc, toxic symptoms, B. not spreading. *Guineas*, by rubbing into skin, death with haemorrhagic serous inflammation. *Mice* die by sbc. injec, 1-10d. B. in organs. *Hab.* isolated from acne contagiosa in horses.

    2. *Grow well on ordinary media.*

        a. *Grow at ordinary room temperatures.*

        * *Milk coagulated, acid.*

        143. *Bact. sputigenes tenuis* (Pansini).

    B. small, variable length, in 2's and chains, in body with a capsule. *G cs* slightly spdg, raised ., circular, yl, concentric, radially striped. *G st* (1) beaded (2) ylsh. *P.* ylsh, flat.

    Path. mice, guineas, refractory to ordinary quantities. *Rabbits*, 1/2-1 c.c. sbc, death, septicaemia slight local effect, hemorrhage in peritoneum, spleen enlarged. *Hab.* assoc. with advanced phthisis and catarrhal pneumonia.

    Bact. pyogenes minutissimus (Kruse) from pus in man, from descriptions not diff. from above.

    ** *Milk becomes thick, viscous.*

    144. *Bact. viscosus lactis* (Adametz).

    B. 1.1-1.3;1.2-1.7—threads with a capsule. *G cs* (2) wh, slimy, erose. *A sl* dirty wh, slimy. *M* 5-10d, becomes thick, viscous → pep and clear, no special odor. *Hab.* milk.

        b. *Growth only above 27° C.*

    145. *Bact. sanguinis typhi* (Sternberg).

    B. typhi like 0.5-0.8: 1-2.5. *Acs* bl, gr, transl, irreg →, dry. *P.* invisible growth. *M co—. Path* ⌐? *Hab.* isolated from blood of typhoid fever patients.

CLASS VI. GELATIN LIQUEFIED. MOTILE.

I. Grow well on Nutrient Gelatin.

A. Colonies on Gelatin at all stages round with no radiations from their edge.

    1. Gelatin Liquefied rather quickly.

        a. *Gas Generated in Glucose Bouillon* (B. CLOACAE GROUP).

*Milk coagulated.

†Reduce nitrates to nitrites.

146. Bact. cloacae, (Jordan).

B. o.7-1.0: o.8: 1.9 (20° 37°) · , *G cs* (1) round, ylsh, (2) thin, blsh, entire-erose, dark center, clear outer zone, crateriform liq. *G* st liq, napiform. *B*. turbid, memb . *Asl*, porcelain wh. *P*. ylsh. *M* ac. H ı
Co₂ 3

Hab. Water, sewage, etc.

††Nitrates not reduced to nitrites.

Gelatin stab cultures crateriform becoming stratiform.

147. Bact. coadunatus (Wright).

B. medium, short, rounded, pairs, threads, chains. *G cs*. 3-4 d. round, brsh, dense, less than 1 m. m. sunken in liq gel. (b) brsh-brsh gr., center with rough frayed margins and zone of liq. gel. in which are scattered granulations. *G st*. crateriform-stratiform. *A sl* transl, grsh, spdg . *B*. turbid, wh, sed, pel, slightly g tint. *B*. br, viscid, mo, gl, spdg. *M* ac+. In ⊢ (37°—). *Hab.* Water.

Gelatin stab becoming saccate.

148. Bact. multistriatus (Wright).

B. medium, rounded, variable, pairs. *Gcs*. (1) brsh, dense, gran, round-oval (2) (a) round, gr wh, transl disks, 1-2 m.m. (b) d brsh, dense in center, thinner at margins, radiate brsh lines from a central neclcus. *G st* (1) beaded (2) irreg, whsh, gradually sinking into liq. gel + 10 d. saccate. *Asl*, transl. narrow P grsh-creamy, thick, gl, spdg, viscid. *M*, am. *In* + (35°—36°)—. *Hab.* Water.

 b. No Gas generated in glucose bouillon.

 * Milk coagulated.

 † Produce indol.

 149. Bact. Fairmontensis (Wright).

B. medium, short, rounded, pairs-threads. *G cs* (1) round-oval, dense gran, dark grsh br; 3 d cs. surrounded by a zone of liq. gel (2) 2 d. round wh, transl disks, 1-2 m.m. (b) center dark with greenish schimmer thinner towards edge, faint radial lines. *G st*, crateriform, extending to walls in 2-3 d, little growth along needle track. *A sl*, grsh, wh, gl, elev ˒, transl+, gsh _ . *B*. turbid, pel ˒, floc sed +, faint g. *P*. dry gran, elev, spdg. color of *P* + brsh viscid. *Lit M* decolorized. In · (35 —36°)—. *Hab* Water.

 150. Bact. duplicatus (Wright).

Can not be distinguished from description from preceding.

 **Milk not coagulated.

†Growth on potato smooth.

§Liquefaction crateriform-stratiform.

151. Bact. formosus (Ravenel).

B. slender, rounded 7-11 x's breadth. *M* o△. G *cs* (1) round, entire, ylsh. gran. . (2) round, entire ylsh. center, gr. edges, gran, later 3 d. a ylsh br, nucleus+, zone lyl with br. wavy lines+, darker zone with yl. exteriorially, in which are also radial lines. *A sl* wh, limtd, edges edges notched, mo, gl. *Gsl* crateriform + stratiform. *P* mo, wh, spdg. —creamery. *B.* turbid, sed. *Lit M* more blue+, discolored in 10 d. In—. (20°-36°) . . Op. room temp. *Hab.* Water.

§§Liquefaction funnel formed.

152. Bact. stoloniferus (Pohl).

B. 0.8:1.2 G *cs* round, sharp border, *G sl* funnel. *A sl* wh, thick with streaming outgrowths. *P* small pin head cs, spdg. *Lac B* no gas. *Lit gel* red, *M* am, *Lit M*, 24 d. unchanged. In—. Ph—. *Path*—. *Hab.* swamp water.

††Growth on potato rough or folded.

153. Bact. antenniformis (Ravenel).

B. large straight rods, rounded ends, 8-10 x's breadth, singly, actively motile. *Gcs* (1) 12-14 hrs. oval, ylsh, gran ; from poles, fine short projections like antennae of insects disappearing in 36 hrs. (2) 36 hrs. 0.25 m. m. grsh, center orange br with a fringe of wavy lines ; border colorless, parallel filaments, dentate ; liq, crateriform, with a pelicle becoming folded+, 7 d. 6 m. m., circular with entire border. *G sl* crateriform—stratiform 10 d. *A sl* thin, smooth, grsh. *P.* 2 d. in- visible. 3-4 d. spdg., layer thrown into fine folds like herpetic vesi- cles+, putty colored and drier and folds more numerous. *B* . *PR* , slight deepening. *Lit M* decolorized+. watery, ac . In. —. (20°- 36°)+. *Hab.* Water.

C. Gas Production in Glucose Bouillon not stated.

  * Potato cultures redish or pinkish, flesh colored.

  † Gelatin stab, saccate liquefaction.

    154. *Bact. aquatilis communis.*

      (B. punctatus, Zimmermann.)

      (B. liquidus, Frankland.)

      (B. liquefaciens communis, Sternberg.)

    B. 0.6:1.2·5 G *cs* (1) round (2) round, crateriform liq. turbid, edge finely gran, not cilliate. *A sl* transl, gr. *G st* 2d, a large saccate liq, turbid → clear. *P* ylsh br—pinkish, flesh colored, nitrites. Hab. water.

  †† Gelatin stab—stratiform liquefaction.

    155. *Bact. bucalis fortuitus.*

      (B. j. Vignal.)

    B. with square ends, 1.4·3, long. often in pairs, jointed at an angle, Mo ?

40

G cs 48 hrs. small, round wh → 4 5d. liq. *G st* (1) ⟩ (2) 2d small punctiform → 4d. spdg. over entire surface, liq. stratiform. *A sl* small, wh, opaque cs. *B.* turbid, memb. *P* thick spdg. , pinkish. *Hab.* isolated from saliva.

** Potato Cultures yellowish—brownish.

  † Gelatin stab. funnel formed liquefaction.

    156. *Bact. hydrophilus fuscus* (Sanarelli).
        (B. ranicida, Ernst.)
    B. 0.6:1·3 filaments, 37°, . *G cs* round transl.
    A sl gr bl-brsh. thin. *B* turbid, memb. △, *P* yl-br.
    Path sbc.-inmusc, haemorrhagic septicaemia in frogs, salamanders, fish ; also, guineas, rabbits, mice.
    Hab isolated from water, frogs dead of septicaemia.

    157. *Bact. pyogenes foetidus liquefaciens* (Lanz).
    B. variable length, 0.5-0.7 broad, liq +, bad odor. *G cs* not described. *G st* funnel. *A sl* thin, whsh, glassy. *M* Co—. *P* citron yl, gas. *Path rabbits* inven inoc 1 c.c. B, a multiple supurating inflam. of the joints.
    Hab isolated from brain abscess after *Otitis media.*

  †† Gelatin stab, crateriform liquefaction.

    158. *Bact. No. 46* (Conn).
    B. 0.4:0.8, chains (20° 37°) · . *G cs* central nucleus, border crenate, clear outer zone.
    A sl thin, wh-ylsh. *P* ylsh, br. *M* Co. ; , am-alk.
    B turbid, ylsh sed. *Hab.* milk.

*** Potato cultures whitish grayish.

  † Milk coagulated.

    ‡ Growth and Liquefaction of Gelatin more rapid than evaporation in stab cultures.
      159. *Bact. delictatulus* (Jordan).
    B. 1;2 (20°-37°) + *G cs* whsh, homo, entire, radiating edge. 2d. dark nucleus, with clear zone of liq. gel. *G st* 2d. funnel, memb. sed. brsh.
    A sl gl. porcelain wh. *P* thin, gr. *M* ac.
    B turbid, sed. scum, nitrites, *Hab.* water.
    ‡‡ Evaporation more rapid than growth and liquetation of gelatin, in stab cultures funnel partly empty.
      160. *Bact. circulans* (Jordan).
    1.0 : 2·5-chains (20°-37°) + *G cs* 2-4d. round, brsh → depressions in liq. gel. *G st* , along line of stab a conical cavity, ppt. in bottom, liq. gel, drying out, leaving cone partly empty.
    A st thin transl layer *P* , color of potato. *B* turbid, no memb. *M* ac ⟍, Co △. Nitrates reduced △ to nitrites. Hab. water.

  †† Milk not coagulated.

    161. *Bact. septicus putidus* (Roger).
    B. like Proteus vulgaris *G cs* round, entire.
    Hab. isolated from a cholera corpse.
    162. *Bact. albus putidus* (Maschek).
    From description not differentiated from above *Hab.* isolated from water.

D. Gas produced in ordinary gelatin or bouillon.

* Grow at 37°C, and more or less pathogenic.
163. *Bact. tachyctonum* (B. Fischer).
   B. middle sized—threads (/ *cs* like cholera. *G st* crateriform → saccate, inemb. *B* memb. gas -. *A sl* brsh. *P* gr br.—rdsh br. *Path* sbc. inper, not too small quantities, septiacemia and bloody odema in mice. gulineas, rabbits—, *Hab* faeces in cholera nostia.
164. *Bact. dubius* (Bleisch).
   Differs from preceeding *P* pale yl. from beginning.
   B no memb. and less pathogenic. *Hab* isolated from faeces.
** Do not grow at 37°C. Water Bacteria.
165. *Bact. gasoformans* (Eisenberg).
   (B. liquefaciens Sternberg.)
   B. small. *G cs* round entire crateriform, spreads rapidly. *G st* saccate, turbid, much gas. *A sl* dirty wh. *P* lylsh. *Hab.*

2. Gelatin liquefied very slowly.
   a. Gas generated in glucose Bouillon.
   * Milk coagulated.
   165. Bact. Kralii (Dyar).
B. Short rounded, o.7: o.8, singly, mot   , *Gel.* liq. ± 30 d ; nitrites. *PR*±. *Lac. Lit.* red → blue. *A sl*, wh, opaque. *Hab* cult. from Kräls laboratory.
   166. Bact. (b) Guillebeau.
B. 1: 1-2. Growth like B. aerogenes. *M co* quickly. *Hab.* isolated from milk.
   ** Milk not coagulated.
   167. Bact. nebulosus (Wright).
B. medium, motile, fla. polar. *G cs* (1) round, dark gran, brsh gr. tint, (2) round, thin, gr, transl, hazy, 3 m.m.; wh center surrounded by a whsh, ring (b) center d.brsh, gran, surrounded by a thin transp zone, *G sl* viscid, whsh, lines a shallow furrow,with short lateral out growths. *A sl* thin, transl stripe. *B.* turbid, sed. *P.* scanty growth, if any. *Lit. M.* decolorized, casein dissolved, alk. *In—*, (35°—36°) ·. *Hab.* Water.
   *** No growth in Milk (?)
   168. Bact. halophilus (Russell).
B. o.7: 1.5—3.5. Grow only in gel, best in sea water gel. *Gcs* round, gr wh, transl. *Gst* liq. \ ; evaporation causes an empty funnel. Cults. alk, much gas. *Hab.* Marine.
   b. No Gas generated in glucose Bouillon.
   * Bouillon cultures with a wrinkled membrane.
   169. Bact. cohaereus (Wright).

B. Medium, short, rounded, pairs—threads, polar fla. *G cs* (1) round—oval, gran, brsh, sharp + d brsh tint in adjacent gel, sometimes morula like. (2) 4-5 d round, elev. grsh, 1·2 m.m.+ thicker, denser, papillate in center. (b) gran, ylsh br in center + sunken in gel, crimpled. *G st* wrinkled , gr wh, lining a furrow in gel. *As l* elev, gr wh, transl, gl. *B* turbid, wrinkled memb. + clear *I* elev, gran, color of potato. *Lit M* decolorized, viscid, co. alk + brsh. In—, $(35^\circ—36^\circ)$. *Hab.* Water.

**Bouillon cultures without a membrane, oxidize nitrites to nitrates.

170. Bact. nitrificans (Burri-Stutzer).

B. 0.5:0.7·1.5, involution forms, stains badly in aqueous analine, *G cs* (1) round, d wh. (2) round, slimy, colorless, which after 8 d. are sunken in liquefied gel., *G st* (2) spdg, colorless—blsh, 2-3 m. m, which begins to sink in the gel, 3 w. crateriform-napiform. B turbid , sed. , whshrdsh. Oxidize nitrites to nitrates. *Hab.* isolated from soil.

c Gas production in Glucose Bouillon not stated.
* Grow on potato.
† Rods scarcely longer than broad, thick ovals.
171. Bact. guttatus (Zimmermann).
B. 0.9:1·1.1. *G cs* (1) brsh center, border bright. (2) Small, round. *G st* (1) (2) *A sl* gr, limtd. *I* slimy, ylsh gr, *Hab.* water.
†† Rods, several times longer than broad.
172 Bact. inunctus (Pohl).
B. 0.8-0.9:3.5 *G cs* entire, round. *G st* (1) lower end, radiating outgrowths (2) thick, gl.
A sl whsh. *I* slimy. *Hab.* water.
** Do not grow on potato.
† Evaporation equals or exceeds liquefaction of gelatin, causing cavities in gelatin.
173 Bact. litoralis (Russell).
B. 2-4 x's breadth, grow slowly. *G cs* (1) 3d. small, brsh. (2) entire, shining—opalescent, gran. liq. 5 8d. Evaporation causes depressed cs. *G st* (1) (2) thin, becoming depressed. *A sl* slimy, wh. *B* turbid, memb. *Hab.* mud bottom Gulf of Naples.
†† Liquefaction equals or exceeds evaporation of Gelatin.
174 Bact. superficialis (Jordan).
B. 1:2.2. *G cs* (b) round, segmented, cracked (2) (a) punctiform transl. (b) round, homo-finely gran, liq ., with a ylsh br opaque center and transl edge, *G st* (1) —, (2) growth almost entirely on surface. *M* 20d. Co.—, ac. .
B. turbid. . No memb. Nitrites—? *Hab.* isolated from sewage.

B. Colonies on Gelatin, with filamentous borders or radiate (B. CENTRIFUGANS GROUP).

1. Gas generated in glucose bouillon.

175. Bact. centrifugans (Wright).

var. a.

B. medium, rounded, pairs—threads, flag, polar. *G cs* (1) round, dark gran, greenish shimmer, soon surrounded by a zone of liq. gel. (2) 24. 48 hrs, round, crateriform, 3 m. m, turbid, floculi in center ; (b) gran, circulating motion, margin fringed with short hairs. *G st* saccate, spdg, pel, greenish below, alk. *A st* transl, gl, grsh thin→, brsh—gsh br. *B* turbid, pel. ' →, brg. tint. *P* thick, spdg, gr—pinkish, sometimes rough gran. surface. *Lit M* Co, decol, am—ac. In :-. Nitrites— 35˚—36°÷. *Hab.* Water.

2. No gas generated in glucose bouillon.

a. Produce indol, grow at 36° C.

176. Bact. fimbriatus (Wright).

B. medium, blunt ends ., short-long forms-chains ; fla. several. *G cs* (1) round, gran, grsh in center, sharp, (2) 2d. rounded, ylsh wh, sunken, 1-2 mm., sometimes surrounded by a clouded liq. zone (b) d brsh gran. center, edge delicate fringe. *G st* napiform, pel. ., irri-descent→ g sh. . *A sl* smooth, d gr, gl., agar→ brsh. *B* turbid, pel. →+ dgsh tint. *P* grsh—l brsh, rough △, spdg. *Lit M* Co+, decol, am. *In*+. Nitrites—. 36°+. *Hab.* Water.

b. Do not produce indol. No Growth at 35° C.

177. Bact. geniculatus (Wright.)

B. medium. pairs—threads *G cs* (1) (a) ylsh, (b) round, sharp, gran ., studded with small plaques or buds. (2) 2d. round, transl. whsh, de-pressed ., (b) 2d. brsh, gran center, thin margin, entire--undulate, liq center rugged+, gran zone with radiating fibrils, greenish shimmer. 3d. grwh, ylsh centers, 3 m.m, crateriform, zone with radiating fibrils. *G st* funnel, air space above, sed whsh—pinkish. *A sl* grsh, gl, limtd. transl—brsh gr. *B.* turbid, floculi in suspension, pel.△, △gsh tint. *P.* thin, visc d, mo, gl, brsh. *Lit M* Co, decol, alk. In—. (35°−36°) —. *Hab.* Water.

3. Gas production in glucose Bouillon not stated.

a. Grow well upon potato.

* Margins of gelatin colonies fibrillous—flocose.

† Growth on agar, dry, dull, tough, becoming rough, warty.

178. Bact. hyalinus (Jordan).

44

B. 1.5: 4— chains.; fac. an. *Gcs* 24 hrs. plainly visible to naked eye, center dark, transl. (b) center coarsely fibrillous with short fibrils radiating from the edge; 2d. 15 m.m. *G st* funnel—saccate, 8 d highly tenacious scum. *Asl* dry, dull, tough, gr.—+ rough warty. *B.* turbid, scum. *P* as in *Asl. M.* 2d Co, ac. Nitrites. *Hab.* Water.

 †† Growth on agar, thin, smooth glistening.

 179. Bact. pestifer (Frankland).

B. 1: 2.3—filaments, grow slowly at room temp. *G cs* (1) irreg. (2) center smooth, margins of wavy bundles. *Asl* gl, transl. *P.* thick, irreg., flesh colored. *Hab.* isolated from air.

Bact No. 9 (Pansini) sputum; B. pneumonicus agilis (Flügge). Indt from 179. Comp. B. vermicularis (Frankland).

 ** Gelatin colonies rosulate.

 180. Bact. meningitidis aerogenes (Centanni).

B. 0.35: 2-2.5, seldom in threads. *G cs* daisy shaped (ganseblumchen), rosulate. *G st* liq. ₃, gas+. *Asl* porcelain wh. *B* turbid. *P* grsh yl, uneven, rough. *Path.* Rabbit's subdural injec, death in few hours to days or weeks with progressive palsy, emaciation and lung complications, hyperemia of meninges, etc. *Hab.* isolated from two cases of menigitis.

 b. Little or no growth on potato.

 * Gelatin stab culture becomes an empty funnel from evaporation of slowly liquefying gelatin. Rods short.

 181. Bact. devorans (Zimmermann).

B. 0.74: 0.9—1.2 solitary, pairs-chains. *G cs* (1) small wh. (2) round, wh, gran—filamentous yl gr, margin fringed. *G st* (1) filiform with a bubble above under which is a whsh growth, funnel forms without visible liquid, with growth along walls of the funnel. *Asl.* thin, gr. spdg. *Hab.* Water.

 ** Gelatin stab cultures not characterized as above. Rods long, slender.

 182. Bact. aquatilis (Frankland).

B. 2.5—filaments 17.0. Grow very slowly in usual media. *Gcs* (2) center ylsh br, from which twisted ylbr filaments are given off. *Gst* (1) scarcely visible at first, liq begins later, then progresses more rapidly (2) small ylsh colony. *A sl* gl, ylsh, limtd. *P* scarcely no growth or faint ylsh line only.

C. Colonies on gelatin, erose—lobed, Coli like (B. DIFFUSUS GROUP.

1. Growth on Potato.
  a. Potato growth grayish—yellowish.
      *Chromatic grains in the interior of the rods.
      183. Bact. Trambustii (Trambusti-Galeotti.)
B. variable, size 3·5, 37°+, chromatic granules in the interior of rods.
G cs irreg edged, gr, surrounded by a zone of liq gel. *A cs* star shaped
with broad radiating outgrowths. *B* not turbid, memb+. *A sl* gr.
P gr, raised dry cs. *Hab.* Water.
      *Not characterized as above.
      184. Bact. diffusus (Frankland).
B. o.5:1.7—threads. *G cs* (1) round, edge erose, gran. (2) thin blshg,
spdg, (b) gran, erose edged, nucleus+ a pale bl, irreg edged outer
zone. *G sl* only on surface thin gl grsh yl, which slowly liquefies the
gel. *A sl* thin, gl, lyl—creamy. *B*. turbid, sed gshyl, with floculi on
surface. *P* thin, gshyl, gshyl, smooth, gl. Nitrites△. *Hab.* earth.
  b. Potato growth yellowish brown (characters coli like).
      185. Bact. sulcatus liquefaciens (Flügge).
B. middle sized. *G cs* (1) small, round, ylsh, (2) large spdg, transl, incised, marmorated, liq . *A sl* transl, gr. *P* ylsh br. *Hab.* Water.
Compare B. coli communis No. 10.
  c. Potato growth brownish.
      186. Bact. delabens (Wright).
B. small, rounded, short and long forms. *G cs* (1) (b) round—irreg,
gran , brsh yl, (2) 2d. thin, transl, gl, wavy-irreg, grsh center (b)
thin, transl, marmorated, brsh center+, sinking in gel, liq ﹄. *Gsl* gr
wh stripe which sinks into gel. *G sl* (1) beaded, brsh gr, (2) thin, wh,
irreg, slowly sinks into liq. gel. *A sl* whsh, transl, gl, spdg △, agar+,
gsh. *B* turbid, pel , gsh tint . *P* brsh, viscid, thick, spdg. *Lit*
M decol, Co—, alk, tough memb. In ﹄, Nitrites—, (35°-36°)—,
Hab. Water.
2. No growth on potato.
      187. Bact. Havaniensis liquefaciens (Sternberg).
B. o.8: 1.2-5-filaments. (20°—37°)+. *G cs* (1) (a) 24 hrs., round,
milky with transp. zone of liq. gel , with irreg. margins. *G st* saccate,
turbid+ clear. *Asl* thin, l brsh. *Path.* rabbits—. *Hab.* surface of
human body.
D. Colonies on gelatin, irregular—fragmentary.
      188. Bact. leporis lethalis (Sternberg).
B. o.5: 1.3-filaments. *G cs* (1) round, transl, lyl. (2) transp, resemble
small fragments of broken glass, later liq occurs rapidly. *G st* funnel.

Asl thin, transl, gl, wh. *P.* thin, spdg, lylsh. *Path. Rabbits.* 1-3 c.c. inper, 2-3 hrs, Somnolent condition, drooping head, death from toxaemia. *Hab.* Intestines of man ill of yellow fever.

II. Grow poorly on ordinary nutrient gelatin unless urea is added.

189. Bact. (Urobacillus) Maddoxi (Miguel).

(Bact. Schützenbergii).

B. o.5: 1.o, usually in pairs. *G cs* small, transl, milky, liq + ; with urea in gel., cs. surrounded by a zone of crystals. *G st* 20° crateriform. *Asl* 28°—30°, whsh, with a slight gsh tint. B. 24 hrs. turbid, memb. *Hab.* fermenting urine.

CLASS VII. GELATIN LIQUEFIED. NON-MOTILE.

I. Gelatin liquefied very slowly or merely softened.

A. Stained by Gram's method (SCHWEINROTLAUF GROUP).

190. Bact. rhusiopathiae suis (Kitt).

(B. of hog erysipelas—Schweinrotlauf, Rouget).

B. very small, slender, bent or curved—threads. o.2: o.6·1.8. *G cs* (2) veily. (b) fine interwoven threads. *G st* (1) gr., cloudy, radiating outgrowths, after some time the gel. is softened. *Asl* a delicate layer. *B.* turbid → gr. wh. sed. *P.* no growth. In—. *Path.* mice, wh. rats, pigeons, death 3-4 d. Septicaemia, B. in blood and leucocytes, mice die in sitting posture with eyes sealed with secretions. *Hab.* Associated with above diseases.

191. Bact. murisepticus (Flügge).

(B. of mouse septicaemia, Koch).

Perhaps identical with the preceding.

B. Not Stained by Gram's Method. (GLANDERS GROUP.)

1. Pathogenic Bacilli.

a. Grow well and even best at body temperature.

192. Bact. mallei.

(Rotzbacillus-Glanders).

B. small, slender bent, o.25·o.4: 1.5-3, coccoid elements, stains .. Grows best on glyc A, *B S*◠. *A sl* 24-48 hrs, 37° whsh, transl, watery cs. *G cs* 20°-25°, after weeks gel. begins to soften and form a small funnel. Grows with acid reaction of medium. *P* ylsh—rdsh br, pleomorphic anthrax like threads→, swollen. In±. *Path.* rabbits but slightly affected, white—gray mice immune, *guineas*, male, inper 1-2 c.c. cult, death 12-15d, testicles swollen and redened, tubercles on tunica vaginalis, supurating organ contain the *B* from which pure cultures can be made. *Hab.* secretions of ulcers and tubercles in glanders, in man, horses, asses, cats, dogs ⌐, sheep, goats and rarely swine; cattle and birds immune.

    b. Do not grow at body temperature.

    193. Bact. salmonica (Emmerich-Weibel).

    (B. der Forellenseuche).

B. short rods, 20° +37-⁰-, *G cs* whgr—+ brsh, gl, cholera like, with rosulate markings. *G st* (1) beaded cs, then air bubbles, with slow liquefaction, and production of an open canal, mostly air. *B* clear, sed. *A sl* thin, mo. gl, gyl-brsh. *P* no development. *Path.* to trout by inoc. of cults. *Hab*, isolated from a disease of trout.

    2 Non-Pathogenic.

        194. *Bact. vermiculosus* (Zimmermann).

          B. 0.85:1.5, capsules, 37° + *G cs* (1) round, gr, gran, (2) spdg, lobed, marmorated, liq △. *A sl* mo opalescent. *P* +, yl gr, gl. Hab. water.

        195. *Bact. incanus* (Pohl-Dyar).

          B. 0.4-05: 0.6-1, chains. *Gel.* liq △. *A sl* transl, wh, irreg, streaked. *P R* ±, *M* Co—, *Lac. Lit* no acid.

        Hab isolated from leaf of Sarracenia purpurea.

II. Gelatin Liquefied Rather Quickly.

  A. Stained by Gram's Method.

    196 Bact. orchiticus (Kutzcher).

B. like glanders. *G cs* like old cholera cs.

A sl dense, wh. *Blood serum*, orange pigment.

B turbid, no sed.

Path guineas, inper, death 4-5d; enlargement of testicles, tubercles on diaphragm, testicles, etc. ; with larger doses the peritoneum is affected as above.

Mice sbc. death 4-7d. abscess, hemorrhagic odema; inper ylsh tubercles on peritoneum.

Rabbits nearly immuue.

Diff from No. 192 by pseudo tuberculous lesions in smaller animals.

Hab. isolated from nasal secretions of a glandered horse.

    197 Bact. pneumonicus liquefaciens (Arloing).

B. short rods—coccoid. *P* whsh br.

Cult. characters not fully described.

Path guineas, rabbits (dogs but little affected) by inper inoc death.

Hab isolated from exudate in lung plague of cattle.

  B Not Stained by Gram's Method.

    1 Potato growth, white or color of potato.

a Potato growth dry irregular surface, grow best at 37· C.

198 Bact. varicosus conjunctivæ (Gombert).

B. 1.o : 2-8, short often constricted in middle. 22° op. 37°. *A cs* 37° 4d 4m.m. with minute thorny projections, opaque nucleus with a ylsh gran transp zone, with twisted bent tapering off shoots.

G st (1) $\diagdown$ (2) circular, flat, grsh wh, center, liq extends gradually downward. *A sl* thin, wh, dry, adherent films. *P* dry, wh, spdg, irreg surface, fringed margin → rdsh br. *Path*. rabbits, inflamation of cornea. *Hab* isolated from healthy conjunctival sac of man.

b Potato groth scarcely visible. Grow well at 20°C.

* Generate gas in nutrient gelatin and glucose bouillon.

199 Bact. aromaticus (Pammel).

B. o 3 o.45 : o.9-1.2 rounded ends 35° . *G st* crateriform—saccate, gas. *A sl* wh, spdg. *M* Co ᾽ , ac, pep. $\triangle$.

Bs dwh, spdg, liq. *P* , invisible unless moist where there form ylsh wh, cs

Hab. Isolated from cheese.

** No gas in nutrient gelatin (?)

200 Bact. nubilus (Frankland).

B. slender, o.3 : 3.0—threads, more or less bent or curved. *G cs* like B. *murisepticus A sl* thin, opalescent. *B*. turbid, sed, pel. *P* scarcely visible growth, Nitrites . *Hab*. water.

2. Potato growth yellowish-brownish.

a. Gelatin stab cultures, saccate liquefaction.

201. Bact. aquatilis radiatus.

(Bact. radiatus aquatilis, Zim.)

B. o.6:1.2 2.5—longer *G cs* surrounded by a delicate "stahlenkranz." *G st* liq+, saccate, turbid, memb. *A sl* gr, transl. *P* ylsh.

a. Gelatin stab cultures, crateriform liquefaction, becoming stratiform.

202. Bact. flexuosus (Wright).

B. medium, thick, rounded—filaments, chains *Mot?* *G cs* (1) round—oval, brsh, gran. (2) 2d. whsh, irreg. clumps (b) twisted strands dense in center, edge irreg. *G st* crateriform +, stratiform, alk.

A sl transl, grsh, limtd, the agar + gsh . *B* turbid + gsh .

P brsh, viscid, gl, uneven. *Lit M* Co. ac. In ? 35-°36°)—, Gl, B. gas—, *Hab*. Water.

3. Potato growth pinkish.

203. Bact. buccalis fortuitus. See Bact. No. 155.

(B. j. vignal).

4. Color of potato growth not stated.

204. Bact. convolutus (Wright).

B. large, long, singly-pairs, twisted chains-threads *G cs* (1) round, dark, gran, in several days cs sink in slowly liq. gel. (2) 2d. round, gsh wh, transl, wooly looking, opalescent△, 2-4 m m. (b) segmented-fissured, edge irreg. darker center. *G st* crateriform, liq ﹨, alk. *A sl* transl, grsn, limtd. agar—+ brshg.

B. turbid, pel—+, gsh. *P* elev. *Lit M.* alk, Co—, In—, *G l B* no gas. (35°36°)— *Hab.* Water.

CLASS VIII. GELATIN COLONIES AMEBOID, MOTILE, GELATIN LIQUEFIED; STAINED BY GRAM'S METHOD (PROTEUS GROUP).

I. Gelatin Colonies Typical of the Group, Ameboid, Cochleate.

A. Agar smear cultures smooth.

1. Potato Cultures white, gray—yellowish, not distinctly brown.

a. Milk coagulated.

205. Bact. (Proteus) vulgaris (Hauser).

B. o.6:1.2·4, threads-chains in flocose arrangement. *G cs* 6-8 hrs, small depressions which contain liq gel and grsh wh. masses of B.

(b) from edge ameboid processes. *G st* saccate.

A sl slimy, mo, gl, transl. *M* Co. 2-3d ac △ —+ ylsh.

P ⌐ ylsh wh, raised. *Albuminous fluids* putrefactive odor, alk. *G l B* gas, Sac. B. gas+. Lac. B. gas—, H 2 H_2S +, In +, urea to $(NH_4)_2$ Co_3. Co_2 1

Path not properly pathogenic for smaller animals.

Filtered cults. in quantity, toxaemia.

Hab. commonly found in putrefying fluids, water etc.

206 Bact. (Proteus) mirabilis (Hauser).

Var. of Bact. vulgaris.

Morph. and in cults. as above, may liquefy gel a little more slowly. Deep cs in gel cochleate, zoogloea.

Hab. putrefying fluids, etc.

Bact. No. 7 (Pansini).

207. From descriptions indistinguishable from Bact. vulgaris.

Path (rabbits, guineas).

Hab. isolated from sputum of consumptives.

b Milk not coagulated.

208 Bact. (Proteus) sulfureus (Holschewnikoff).

B. and cults. to all appearances identical with B. vulgaris. *M* remains unaltered, but gradually becomes peptonized without coagulation, and of a ylsh color. H_2S on cooked egg. *Hab.* water.

  2 Potato cultures brownish, cause septicaemia in mice.

      209 Bact. (Proteus) septicus (Babes).

B. middle sized, o.4 thick, length variable, comma forms—threads. *G cs* B. vulgaris. *P* elev, bright br. *A sl* a reticulatedlayer.

Path mice, sbc, death, septicaemia, B in blood ; rabbits $\triangle$. *Hab* isolated from organs of a child dead with septicaemia symptoms.

B Agar smear cultures, crumpled.

     210 Bact. albus cadaveris (Strassmann-Strecker).

B o.75 : 2.5—filaments. *G cs* ameboid, liq $+$, bad odors. *A sl* crumpled layer. *P* thin, wh yl, gran. *Path* mice, guineas die of toxic symptoms with comparatively small doses.

Hab isolated from the blood of a 4 days old cadaver.

II. Gelatin Colonies Cilliate-Radiate. Related to B. Centrifugans Nos. 175-182 Stained by Gram's Method.

    A. Pathogenic to the smaller animals.

     211. Bact. dysenteriae liquefaciens (Ogata).

B. slender rods, mostly in 2's *G cs* with short radiations.

Gst funnel, sed. turbidity.

Path mice, sbc. o.1 c.c, subcutaneous oedema.

Guineas, sbc, oedema, gray knots liver, spleen and large intestines, hemorrhagic infiltration of large intestines.

Hab isolated from cases of Japanese dysentery.

    B. Not Pathogenic to the smaller animals.

     212. Bact. No. 9 Pansini.

B. variable length. *G cs* edged with radiating filaments. *A sl* gr, transp, stringy. *P* yl, potato colored green, bad odor. *Path*—. *Hab.* isolated from sputum of consumptive.

CLASS IX. GELATIN COLONIES AMEBOID, ETC. GELATIN LIQUE- FIED. NOT STAINED BY GRAM'S METHOD OR INDETERMINATE.

I. Potato Growth Whitish—Grayish.

    A. Gelatin Liquefied rather quickly-pathogenic.

     213. Bact. murisepticus pleomorphus (Karlinski.)

B. variable, coccoid-oval rods-spiralloid-Proteus forms. *G cs* like B. vulgaris and mirabilis.

A cs like B. vulgaris, *P* whgr, homo, spdg.

Path mice, sbc, death 24 hrs, B. In blood.

B. Gelatin liquefied slowly-non-pathogenic.

214. Bact. larvicida (Dyar).

B. 0.8:1.0. Gel liq 14-22d. *G cs* Proteus like. *M Co·+*, Nitrites. *A sl* thin, transl. *PR±*. *Lac. Lit.* red→ blue. From silk larvae of *Clisiocampa fragilis.*

215. Bact. dendriticus (Lustig).

B. 0.5-0.8:0.8-2.0 *G cs* large, raised, wh, mo, gl, with 8-10 branches, liq. very slowly.

G st (1) beaded wh. (2) a semi-spherical bead.

A sl thin, irridescent, *B* memb, adhering strongly to the tube. *P* wh, mo. *Hab.* Water.

CLASS X. GELATIN COLONIES AMEBOID, ETC.

Non-Liquefying.

Motile.

I. Little tendency to develop in the depth of the Gelatin, especially in Gelatin Stab Cultures.

216 Bact. (Proteus) Zopfii (Kurth).

B. like B mirabilis No. 206 →+ threads—coccoid elements. *G cs* generally grow just beneath the surface, branching zoogloea, radiate, filamentous, variable.

G st little growth in depth, but in upper portion of line of puncture a radiately filamentous growth as in plate colonies. 37° △ 20°+. *In* —? (*Lac Lit* blue. *P. R.* ±, nitrites—) Dyar. *Path* (?) See Bact. vulgaris.

Hab. isolated from intestines of fowl.

217 *Bact (Proteus) Zenkeri* (Hauser).
Indistinguishable from and probably identical with the preceding.
217a *Bact allantoides* (L. Klein).
Indistinguishable from and probably identical with the preceding.

II. Decided growth along the entire length of Needle in Gelatin Stab Cultures.

A Growth on Potato very slow and scanty, light becoming yellowish, never brownish. Probably not pathogenic.

218 Bact. arborescens non-liquefaciens (Ravenel).

B. slender rods, 7-13 x's breadth, singly-chains of several elements. Mot ⌒.

G cs 48 hrs. blsh, indistinct cloudy dots easily overlooked, resemble cs of *B ramosus*, but less distinct and finer, *i. e.* radiate, filamentous, branched.

G st (1) 5d fine outgrowths, becoming beaded below. (2) 6d 3m.m. irreg, thicker at center, wh, concentric.

A st faint colorless line with lines of watery colonies on each side, *B* turbid ⟍ → clear. *P R* ⏤ +. *Lit. M.* Co. 1od. flocculi, ac, decol. *GlB* gas—, In—, op. 35°–36°. *Hab* earth 2 feet deep.

B. Potato growth brownish, abundant, strongly pathogenic, mice, rabbits.

219   Bact. (Proteus) lethalis (Babes).

B. 0.8 : 1.5, thick short flask shaped rods—filaments. *G cs* raised, whsh, transl, later, outgrowths which branch on the surface. *A sl* thick, opaque ylsh layer. *P* brsh growth.

Path, mice, rabbits, sbc, death 4d. local odema, septicaemia, enteritis, peritonitis, guineas   .

Hab. isolated from lung gangreene in man.

Non-Motile.

220   Bact. multipendienlatus (Flügge).

B. long, slender. *G cs* irreg outgrowths of zoologea masses to coli like. *G st* (2) short radiating outgrowths.

P. smooth dirty yl.

PIGMENT, CHROMOGENIC, BACTERIA.

CLASS XI. PIGMENT YELLOWISH, GELATIN LIQUEFIED, MOTILE.

I.   Gelatin colonies conglomerate—warty.

A.   Rapidly liquefy the gelatin.

221.   Bact. ochraceus (Zimmermann).

B. 0.7: 1.2-4.5—threads *G cs* granuled ylbr→ beset with warts. *G st* funnel with ylsh sed, gel *G* ylsh. *Asl* thin, ochre yl, *P.* thin, oche yl. *Hab.* Water.

B.   Slowly liquefy the gelatin ; grow feebly in gelatin.

222.   Bact. citreus.

Ascobacillus citreus (Unna-Tommasoli).

B. 0.3: 1.3 singly or in bundles. *G cs* conglomerations of small spheres. *Gst* (1) in funnel small flakes.   (2) a slimy citron yl. layer. *Asl* abundant with honey drop like protuberances. *P.* slimy, citron yl→ 2 w. gsh yl. *Hab.* isolated from surface of human body in eczema.

II.   Gelatin colonies simple, not conglomerate—warty.

A.   Liquefaction of the gelatin near the surface in gelatin stab cultures, viz. crateriform—napiform.

1.   Milk coagulated.

223. Bact. annulatus (Wright). •

B. small, rounded, single, pairs—longer forms. *G cs* (2) 3-4 d, 2-3 m.m., round, saccate liq, ylsh center, edge indistinct $\searrow$ fringed. *G st* crateriform, ylsh, floc, $\triangle$ pel. *Asl* yl, transl, gl. *B* turbid, yl pel. *Lit M.* Co, decol. *In* $\searrow$. (35°—36°)—. *Hab.* Water.

    2. Milk not coagulated.

       a. Produce indol.

          224. Bact. pullulans (Wright).

B. small, short, rounded, pairs. *G cs* 2-3 d, 2 m.m. ylsh gr, elev $\searrow$, transl. (b) ylsh center, colorless margins sausage shaped granules on lower side of colony→ surrounded by a zone or liq. gel. *G st* nail, raise→ napiform liq. *Asl* yl, transl. *GlB* no gas. *Nitrites*—, (35°— 36°) ‡. *Hab.* Water.

       b. Do not produced indol.

          225. Bact. aurescens (Ravenel).

B. short, spindle shaped 2-3 x's breadth, singly. Mo. $\searrow$. *G cs* (1) +. (2) minute whsh points (b) gran. brsh. entire; 4 d (2) 0.5 ylsh br. homo, entire *G st* (1) $\triangle$ (2) yl. button→ ɪ m.m. by the 4th d.; 18-20 m.m. crateriform dry depression. *Asl* thin, ylsh, lim td. + golden yl.

P yl, thick, mo, spdg→ orange yl.

B turbid , dense floc. sed. *Lit M* reduced, no acid.

G lB no gas. (20°-36°)⊤. *Hab.* soil.

    3. Milk coagulation not stated.

          226. Bact. cuticularis (Tils).

B. 0.3-0.5:2-3-threads. *G cs* coli like (1) irreg.—entire, br. (2) middle br, edge colorless.

G st (1) _(2) on surface a yl membrane. *P* $\searrow$, slimy yl. *Hab.* Water.

B. Liquefaction of the Gelatin along the entire length of the needle in stab cultures, viz. funnel formed-saccate.

    1. Gelatin liquefied very slowly, potato culture crumpled.

       227. Back. tremelloides (Tils). •

B. 0.25:1, *G cs* (2) raised, which later becomes spreading.

Gst liq very slowly, equal growth along stab.

P coarse, gran, crumpled, ylsh. *Hab.* Water.

    2. Gelatin Liquefied rapidly.

       228. Bact. of liver abscess (Korn).

B. 0.4-0.6:2-5, rounded, chains, Gm-, Mo. $\triangle$. No Fla-dem. (20°-37°)+ *G cs* round, ylsh, gran, surrounded by a zons of liq. gel, with strong pigment formation. *G st* funnel, ochre yl, memb, bright yl sed. *G sl* raised, gl. ylsh. *P*. elev gl, lyl-brsh yl. *M.* Co.—, alk → oily, $\triangle$ stringy. *Hab.* isolated from a case of liver abscess.

III. Gelatin Colonies Ameboid, Proteus Like. Zoogloea Masses.

229. Bact. dianthi (Arthur-Bolley).

B. oval eliptical, single-rarely united, 0.9-1.0:1-2, in rich fluid media more united forming short filaments; afterwards forming zoogloea. *Neutral G cs* zoogloea bodies with surrounding irreg. area of coalescing forms which appear as large viscid drops. *Acid G cs* zooglea make nearly entire body of colony with more irregular outline and a lobed wrinkled appearance to the surface, color 1 orange. *G sl* smooth, limtd, 1 cream→ rough, tuberculate, 1 orange. *G st* (1) (2) arborescent feathery. *A sl* ylsh. *P.* ylsh. *Hab* Assoc. with bacteriosis of carnations.

IV. Gelatin Colonies not Described.

A Milk Coagulated, at least on boiling.

    230 *Bact. Hudsonii* (Dyar).

        B. 0.5-0.6 : 0,7-1.5, singly-pairs. *P.* thin, transl ; ocherous-orange. *Glyc.* A thick mustard colored.

        Lac. Lit reddened, variable. *P R* ±. nitrites—. *Hab.* air.

B Milk not Coagulated.

  1 Reduce nitrates to nitrites.

    231 *Bact. theta* (Dyar).

        B. 0.5-0.7:1-1.3, pairs, motility (?) *A sl* transl, ocherous. *Cel* liq. △, 2id. *Lac. Lit.* blue. *B* memb. *P* thick, spdg, gl, br, ocherous. *P R* —. *Hab.* air.

  2 Do not reduce nitrates to nitrites.

    232 *Bact. gamma* (Dyar).

        B. 0.5-0.7 : 1-1.5, mo. +. *A sl* spdg, transl, ocherous, concentrically marked, comes off in pieces under needle. *B* thick memb. *M* pep ∠ *Lac. Lit.* blue, reduced. *P R* ± *Hab.* air.

CLASS XII. PIGMENT YELLOWISH, GELATIN LIQUEFIED, Non-Motile.

A Growth on Potato.

  1 Growth on Potato yellowish.

    a Gelatin Colonies, radiate filamentous, branched, arborescent.

    233 Bact. arborescens (Frankland).

B. 0.5 : 2.5—filaments *G cs* as above designated, very characteristic, center ylsh, border iridescent, liq △. *G st* (1) turbidity (2) thin iridescent layer, later ylsh sed. *A sl* and *P.* deep orange.

Pigment insol in water, alcohol sol. *B* turbid. memb—, ylsh sed, nitrites—, (M. Co—, Dyar) *Hab.* water.

    b Gelatin Colonies not characterized as above.

      * Milk coagulated.

† Lactose litmus milk redened.

234 Bact. oxylacticus (Dyar).

B. 1-1.3 : 1.7-2.5, singly—chains. *G el.* liq quickly. *P* gl. transp. watery, marked with opaque wh spots. *A sl* wh, slightly ocherous tint, *P R* ±, nitrites—, *Hab*, from Krals Lab.

†† Lactose litmus milk, rendered alkaline or unchanged · in reaction.

§ Gelatin liquefied rapidly.

235 Bact. desidiosus (Wright).

B. small, short, rounded, pairs-clumps. *G cs* 3d. 1m.m. ylsh br, irreg clumps in liq gel. (2b) dense, gran, brsh yl, often darker in center, irreg and broken. *G st* liq, fusiforms bubble at top. *A sl* gl, brsh yl, transl, spdg △. *P* brsh yl, mo, elev, rough. *Lit M* viscid clot, yl, yl br serum, △ alk-am. *G lB* no gas. *In* △ ? (35°—36°)—. *Hab* water.

§§ Gelatin liquefied very slowly, milk coagulated very slowly.

236 Bact. Fischeri (Beyerinck-Dyar).

B. 0.5-0.6 : 0.6-1.2, singly. Gel. liq ., 28d, on solid media ylsh, but pigment forms slowly. *M* Co slowly often 28d, *nitrites* ., *P R* ⊢, *Lac Lit* ±. *Hab*. from Krals Lab.

**Milk not Coagulated.

†Litmus Milk redened, rendered acid. .

§Nitrates reduced to nitrites.

237. Bact. eta (Dyar).

B. 0.5:0.7-1.0, singly, *G el*, liq△. *M* Co-. *PR*±. Growth on solid media yl, viscous. *Hab*, air.

§§Nitrates not reduced to nitrites.

238. Bact. fulvus (Zimmerman-Dyar).

B. 0.8:0.9-1.3, singly, pairs-short chains. Op. 30° liq . On all media gamboge yl. *G st* (1)+(2) round, convex, ylsh. *P*+, gl, ylsh, (M· Co—, nitrites—, Lac. Lit. reddened) Dyar. *Hab*. Water, etc.

††Litmus Milk, reaction alkaline or unchanged.

§Agar Smear Cultures, redish brown, shading to yellowish.

239. Bact. rubidus (Eisenberg-Dyar.) See 266.

§§Agar Smear Cultures, yellowish-orange.

Gelatin Liquefied quickly (i.e., within a few days).

240. Bact. caudatus (Wright).

B. small, slender, conical ends, pairs-threads. *G cs* 3-4d. 1-2 m.m, yl, transl, smooth-wavy. (2b) ylsh center, radiating ＼, light periphery.

G st (1) villous (2) crateriform, ylsh, floc.

A sl transl, yl, gl, spdg . *B.* turbid, ylsh, sed. *P* dyl, elev, spdg,
uneven. *Lit M* decol . *G lB.* no gas. *In ?* ($35°$-$36°$)— *Hab.*
Water.

Gelatin Liquefied very slowly (i.e., within 2-4 weeks).

241. Bact. fuscus liquefaciens (Dyar).

B. 0.5-06:1-2. Singly-short chains. Gel. liq ͻ, 14·50d. *A sl* bright
orange, forming a crusty skin, nitrites ͻ. *Lit M* blue.

Hab. From Kral's Lab., air.

242. Bact. helvolus (Zimmermann-Dyar).

B. 0.5:1.5-2.5-4-5. Op. 25° *G st* (1)ͻ (2) beaded, aerogenes like—+
spdg, Naples yellow + crateriform liq.

A sl Naples yellow. *P* yl + gsh△.

(*G el.* liq. 30:40d ; *nitrites* ͻ, *M* Co --, *Lit M* blue) Dyar.

Hab. Water, air.

*** Milk coagulation not stated, growth on liquefied Gelatin crumpled.

† Gelatin colonies moruloid.

243 *Bact. plicatus* (Zimmermann).
  B. small, slender, in 2's or short chains; op. 20°c. *G st* (1) beaded, ylsh
  wh, (2) wash yl, crumpled, memb. which gradually sinks. *P* thin, dry,
  ylsh gr. *Hab.* water.

†† Gelatin Colonies punctiform.

244 *Bact. citreus cadaveris* (Strassmann-Strecker).
  B. 0.6-0.9, often in chains, liq . . *G cs* punctiform. *G st* air bubble above,
  under this a ylsh layer, under this clear canal of liq gel with ylsh sed. H₂S
  odor. *P* citron yl, dry. *Path* negative. *Hab.* isolated from human
  corpse 50 hrs. after death.

2. Growth on potato brownish or becoming so.

  a. Broad bacilli, scarely longer than broad.

    245. Bact. buccalis minutus (Sternberg).

      (B. g. Vignal).

B. Rods scarcely longer than broad, 0.5-1.0. *G cs* (2) 48 hrs, round,
elev. △. yl.—+ surrounded by liq. gel. *G st* 48 hrs. (1) ylsh wh
beaded. (2) 3 m.m. ylsh + cup shaped liq., in 6 d. a small funnel, in
12 d. liq. complete. *Asl* golden yl. cs. easily removed with needle.
B. turbid iridescent pel. *P.* thin yl + brsh. *Hab.* isolated from
saliva of healthy persons.

  b. Bacilli minute slender.

    246. Bact. pseudo conjunctivitidis (Kartulis).

B. minute, 0.25: 1.0. Cultures produce a canary yl pigment. *G st* nail
with flat yl head, liq. ͻ. *P* .limtd, bright br. *Hab.* isolated from
conjunctival secretions.

247. Bact aeris minutissimus (Bey).

From descriptions indistinguishable from preceeding, except th..t it produces pigment a little less strongly. *Hab.* Isolated from the air.

    B. Do not grow on potato.

       248. Bact, dormitator (Wright).

B. medium, conical ends, variable, long pair-threads. *Gcs* 2d. 1 m.m. ylsh point (b) ylsh, gran ,. rough sharp bulging outlines, some surrounded by a zone of liq. gel. *Gst* funnel, turbid, sed bright yl. *Asl* gl, transl, ylsh. *B.* turbid, ylsh, sed pel . *Lit M* decol, am. *G IB* no gas. *In*—. ($35°$ —$36°$)—. *Hab.* Water.

Class XIII. Pigment Yellowish, Gelatin not Liquefied. Motile.

I. Gelatin Colonies Large Spreading.

  A Chromogenic Function weak, pale yellows.

    249 Bact. subflavus (Zimmermann).

      (Bact. flavescens, Pohl)?

B. 0.8 : 1.5-3. *G cs* (1) round, ylsh wh. (2) punctiform —+ spdg, border irreg. (b) pearly lustre, dirty yl. *G st* (2) thin, ylsh, gr. *A st* pale yl, spdg. —+ darker pale chrome yl.—ochre yl. *P* scanty, clay yl. *Hab.* water.

    B Chromogenic function strong, deep or golden yellows.

      1 The Gelatin acquires a green fluoroescence.

        250 Bact. fluorescens aureus (Zimmermann).

B. 0.75 : 2.0, op. 20°. *G cs* (1) round, gran, lyl. (2) ylsh gr, margins ill-defined, irreg, thickest in the middle. *G st* (1) △ (2) thin layer, ylsh, spdg. *A sl* ochrous—golden yl. *P.* as before. *Hab* water.

      2 Do not cause fluorescence of the Gelatin.

        251 Bact. aureo fllavus.

        (B aureus, Adametz.)

B. 0 5 : 1.5-4. Grow slowly at room temperatures. *G cs* (2) 8d. small wh. points —+ ylsh, opaque, round—eliptical, 1-4m.m. *G st* (1) △ (2) small round cs, crowded to make a dark chrome yl layer. *P.* gl. hemi. sph. cs —+ a chrome yl layer —+ rdsh br in old culture. *Hab.* water, surface of body in eczema.

II. Gelatin Colonies, Aerogenes Like.

    252 Bact. aurantiacus (Frankland).

B. short, thick, variable, often in threads. *G cs* (1) small, round, gran. (2) aerogenes like, opaque, homo, bright orange. *G st* (1) △ (2) gl, orange. *A sl* limtd, orange. *B* clear, orange sed   , thin pel. with orange flecs. *P* thick, gl, limtd, orange red. Nitrites △. *Hab* water.

III. Gelatin Colonies not Described.

253 Bact. campestris (Pammel).
B. 0.37 : 1.8-3 *A sl* cadmium yl. B ylsh wh, memb. Sac B. gas—. Decay of turnips.

CLASS XIV. PIGMENT YELLOWISH, GELATIN NOT LIQUEFIED. NON-MOTILE.

I. Chromogenic Function weak, pale or grayish yellow.

A. Gelatin Colonies beset with thorny outgrowths, when old.
254. Bact. spineferus (Tommasoli).
B. 0.8-1.0:2, bent, often parallel in bundles.
G cs old colonies with thorny outgrowths.
A sl grsh yl. *PR*. yl. *Hab*. isolated from surface of human body.
  B. Gelatin Colonies not characterized as above.
    1. Litmus Milk redened or rendered acid.
      a. Nitrates Reduced to nitrites.
        *Milk Coagulated after a long time by boiling (26d).
        255. Bact. subochraceus (Dyar).
B. 0.7:1.5, singly-short chains. *G cs* (1) round, dusky, ylsh. (2) clear, irreg, △veined, nitrates△. *Lac Lit* red, 35d purple, 9od. blue. *On solid Medir* transl, ocherous tint—l orange. *B* surface memb.△ *PR* deepened. *Hab*. air.
        ** Milk not Coagulated.
        256. Bact. domesticus (Dyar).
B. 0.5:1.0. *G cs* (2) thin, entire, gran, ylsh centers. *On solid media* wh—+ lyl, spdg, slowly. *Lac Lit*. red—+ 5od blue. *Hab*. blue.
        257. Bact. amabilis (Dyar).
B. 0.7:0.8-1.0, singly, chains-masses. *G cs* (2) large, transl, ylsh. *A sl* limtd, wh. with ylsh tint. *P* thin, bright yl. In—, *PR*±. *Hab*. air.
      b. Nitrates not reduced to nitrites.
        258. Bact. lacunatus (Wright).
B. small, short, rounded, pairs-small clumps. *G cs* (1) round, entire, gran△, gr br. (2) 24 hrs, thin, transl, grsh center, very irreg, deeply cleft. (b) areolate—grained cs—+ 4·5 m. m, thin transl, yl haziness about their centers. *G sl* thin, transl, grsh—+ ylsh in center.
B turbid, *P* thin, viscid, dirty brsh.
Lit M Co—, ac—+ brsh. *G lB* no gas. *In*+ (35°-36°)—, *Hab*. Water.

2. Litmus Milk blue, or reaction unchanged.
   a. Nitrates reduced to nitrites.
     259. Bact. javaniensis (Dyar.)
B. short-coccoid, 1·1.2 diam, masses to short chains. *A sl* thick, wh, with an indistinct ylsh tinge. *M* Co—, *PR*± *Hab.* Air.
   b. Nitrates not reduced to nitrites.
     * Cultures on solid media wrinkled.
     260. Bact. fuscus pallidior (Dyar).
B. 0.5-0.7: 1·1·3, singly—chains. *On solid media* pale whsh orange almost pinkish, wrinkled with lobed edges and crusty brittle texture; forming a surface memb. on liquid media. *Lac. Lit.* blue. *Hab.* air.
     ** Cultures on sold media not wrinkled.
     261. Bact. ovalis (Wright).
B. meduim sized, short, rounded, pairs—long forms. *G cs* (1) opaque, gran, entire, brsh. (2) 5-6 d. 1 m.m., round, entire, elev, gl, ylsh, transl,—+ 2 m.m. yl—br yl, gel △ brsh. *Gsl* elev, br yl, limtd, smooth —rugged, gel—+ brsh tint. *Asl* pale yl, thick, gl, limtd ꓳ. *B* clear, sed. *P* viscid, mo. br yl. spdg. *Lit M.* decol, Co— *In* △. (35°— 36°)—. *G lB* no gas. *Hab.* Water.

II. Pigment strongly developed, yellow color decided.

A. Gelatin colonies, compound.
     262. Bact. luteum (List).
B. 1.1: 1.3. op. 30°. *G cs* irreg. flat (b) consisting of many club shaped coarse gran zoogloea masses, orange yl. *M* Co+. *Hab.* Water.
B. Gelatin colonies simple.
   1. Gelatin colonies, dry, granular, stain irregularly like diptheria B.
     263. Bact. striatus flavus (v. Besser).
B. Small, often bent. *Gcs* thick, dry, gran, yl. *Asl* and *P* sulphur yl. *Hab.* Isolated from nasal mucus.
   2. Gelatin colonies not characterized as above.
     a. On agar a wrinkled layer.
     263. Bact. fuscus (Zimmermann) (Dyar).
B 0.6 wide and of variable length. op. 30°. *G cs* (1) round, irreg, gran, gr yl—br. (2) punctiform, brsh yl center, lighter border. *Asl* crumpled chrome yl, thick, *P* chrome yl, friable.
Lac. Lit. blue.
(Nitrites—, *B* memb, PR±) Dyar. Hab. water.
     b. Agar cultures not characterized as above.

   * Uniform growth along needle track in gelatin stab
cultures.

   † Large, long rods.

   265. Bact. constrictus (Zimmermann).

B. 0.75: 1.5-6.5- chains. Grows only at room temps. *G cs* small, gl,
erose edges, Naple yl· *Asl* and *P* yl. *Hab.* Water.

   †† Short rods and coccoid Forms.

   266 Bact. brunneoflavus (Dyar).

B. short rounded, 0.5 : 0.6-1.0, singly-short chains. *B* no memb. Growth
bright orange. *Nitrites* . *P R* ±. *Hab.* from Kral's Lab.

   ** Scant growth along needle track in gelatin stab cultures.

   267 Bact. flavocoriaceus (Eisenberg-Dyar).

B. very small. *G cs* small, round, sulphur yl → irreg edged. *G sl* (1)
  (2) bead.

(*Lit Lac*, blue, *P.* , *M.* Co—, nitrites +, *P R* deepened) Dyar. *Hab*
air, water.

CLASS XV. PIGMENT REDDISH, GELATIN LIQUEFIED. MOTILE.

I. Pigment Bright Carmine or Fueshine Red.

A On Liquefied Gelatin in Stab Cultures the prodution of a Membrane.

   1 Pigment carmine red on agar and potato.

   268 Bact. ruber sardinae (Du Bois Saint Sevrin).

B. short rods, 0.5-0.6, mostly in 2's. *Gel.* liq with a strong memb. of
carmine red color. Pigment weaker in B. or on Agar at 37°. *P.* beau-
tiful carimine red, strong odors of trimethylamine. *Hab* isolated from
sardine oil.

2 Pigment brick red on agar and potato.

   269 Bact. ruber indicus (Koch).

B. small, very short. *G cs* (1) gold yl, erose. (2) edge, incised, torn.
G sl saccate, on surface a fragile red membrane, wh sed.

A sl and *P.* brick red.

Path rabbits, large doses inven, kill, toxemia, etc., weaker than No. 271.
Hab isolated from stomach of an ape.

B No membrane on liquefied gelatin stab cultures.

   1 Pigment granules in the rods.

   270 Bact. ruber aquatilis (Lustig).

B. small, 2-3 x's breadth, within rods red pigment granules. *G cs* erose,
reddish, liq +. *G sl* funnel, in depth no pigment or at 37°. *A sl* and
P. raspberry red. *Hab* water.

2. No Pigment granules in rods.
271. Bact. prodigiosus (Flügge).
B. ruber balticus, Kieler, Wasserbacillus, (Breunig).
B. rosaceus metalloides, Dowdeswell.
B. ruber berolensis (Fränkel).
B. minaceus, Zimm.
B. ruber, Frank.

B. 0.5:0.5—1.0—coccoid forms—+ thread like in△ ac, B ; op. 22°-25 ; 38°-39° no pigment.
G cs (1) round-oval, entire, rdsh-brsh, transl. on border (2) irreg. rough contour, gran, gr br. liq+ when red color of colony appears.
A sl cs whsh—+ rdsh   *G st* saccate liq, rdsh sed.
M Co, ac. pep, ylsh-rdsh   *P* rose rd. mo—+ d rd-purple red, odor of trimethylamine.
B turbid, rdsh, pel△—+, oily consistency.
Pigment. soluable alcohol-ether ; sol becomes orange yl on addition of alkalies, and violet red with acids.
G IB gas production variable, ac—.   H-2S—, *In*△.
Path scarcely so ; cults. toxic. in very large doses.
Hab commonly present on articles of food, particularly starchy materials, also meat, water, etc.

III. Pigment Flesh Colored.
272. Bact. carneus (Tils).
B. 0.5:2.0 *G cs* 2d. round, sharp, crateriform liq, center darker and finally gran, colorless outer zone.
G st funnel, liq+, rose colored sed.
A sl and *P* flesh red pigment.   *Hab* Water.

CLASS XVI. PIGMENT REDDISH, GELATIN LIQUIFIED.
NON-MOTILE.

I. Stained by Gram's Method.
273 Bact. pyocinnabareus (Ferchmin).
B. 0.8 : 2.5—threads. Op. 37° *G cs* irreg erose, gran, liq +.   *G st* funnel, red sed.   *A sl* rdsh, mo.   *P*. ylsh-rdsh.   *B* turbid, red memb, trimethylamine odor.
Path not pyogenic, toxic in large doses.
Hab. isolated from pus.

II. Not Stained by Gram's Method, (Presumibly).
A Milk Coagulated.

1 Litmus milk blue or decolorized, not red, reaction alkaline or neutral.

 a Nitrates reduced to nitrites, peptone rosolic acid, solution unchanged.

  274 Bact. lactis erythrogenes (Hueppe).

  (Baccillus of red milk.)

B. 0.3-0.2 : 1-1.4 *G cs* round, gr yl—yl, rose color to surrounding gel. *G st* (1) (2) thin, whsh ⊣ yl, liq 10—12 liq gel of a pinkish tint. *A sl* ylsn—weak ylsh red.

M Co slowlp ⊣ pep, am—alk, a layer of blood red serum above the precipitated casein, above this the ylsh wh. cream layer. *B* turbid, ylsh tint, disgusting odor. *G lB.* no gas. In ⊣. H_2S. *P R* ·, alk. *Hab.* red milk, water, faeces of a child.

Var. Bact. erythrogenes rugatus (Dyar).

 " Differs from the preceeding in that growth on agar is thin, membranous, wrinkled." Dyar.

  b Nitrates not reduced to nitrites, rosolic acid slowly decolorized.

  275 Bact. exiguus (Wright).

B. small rounded, singly, pairs—clumps. *G cs* 3d. (1) round-oval, gsh center, margin gran. (2) round, pinkish, transl, gl, disks, (b) gran, pink center, lighter towards margins ⊣ liq, salmon pink. *G st* crateriform, pinkish sed. *A sl* thin, mo. pinkish. *B* turbid, wh sed. *P* spdg, mo, gl, rdsh yl. *Lit M* Co, decol, pep . am. *Glb* no gas. In △. (35°—36°) ·. *Hab* water.

2 Litmus milk redened.

 a Growth salmon pink.

  276 Bact. epsilon (Dyar).

B. 0.5 : 0.7 1.0. *Gel.* liq - . *Asl* transl ; pink. *Milk* Co., at least on boiling. *P R* +. *Hab.* air.

b Growth brownish red or dark orange red.

  277 Bact. zeta (Dyar).

B. 0.5 : 0.7-1.0. *Gel.* growth ⌐, liq. does not begin before 10d. *M.* surface a red cream. On solid media as above. Nitrites—, *Hab.* air.

B Milk not coagulated.

1 Litmus milk redened, acid.

  278 Bact. delta (Dyar).

B. 0 5 : 0.8-1.0, singly-short chains. *Gel.* liq, slowly beginning in 21d. *Asl* thin, red ⌐. *M* red on surface. *P* gl, lrd, limtd ⌐. Nitrites —. *P R* ±. *Hab* water.

2 Litmus milk not redened, reaction unchanged or alkaline.

  a Pigment bright red or pinkish.

    279 Bact. haematoides (Wright).

B. medium, blunt, stains irreg. *G cs* (1) round, gran ylsh rd. (2) 5·6d small, gl, elev. _ɔ. vermillion colored disks, entire, gel. liq △ after a long time. *G sl* bright red, rough, gl. *Ac G* +. *Asl* pink, confluent elev. cs. *P* bright red, wrinkled, granular spdg. *Lit M.* decol—, growth rdsh, reaction alk. *G lB* no gas. In—, (35—36°)—. *Hab.* water.

  2. Pigment brownish red.

    280. Bact. rubidus (Eisenberg-Dyar).

B. 0.5·0.7:0.6·1.0, singly·short chains. *G cs* round, gran△, rdsh in center, gel liq, slowly—rather quickly. *A sl* transl, rdsh br, sometimes shading into ylsh-orange. *P* brsh rd. *Hab.* Water, air.

CLASS XVII. PIGMENT REDDISH PINKISH, GELATIN NOT LIQUEFIED.

I. Motile.

  a. Small slender bacilli.

    281. Bact. rubefaciens (Zimmermann).

B. 0.3:07·1.6 *G cs* (1) round, small ylsh·brsh (2) flat, wh·rdsh *G sl* (1) uniform (2) gwh·ylsh→ wine colored. *A sl* thick, gr blue. *P* yl·br with flesh colored edge. *Hab.* Water.

  b Large stout bacilli.

    382. Bact. rubescens (Jordan).

B. 0.9 4, pairs-threads, rods sometimes slightly curved. Mo△, (20ᵛ·37°)꞉.

G cs (2) projecting porcelain wh round, entire→ of a brsh cast. *G sl* (1)_ɔ(2) porcelain wh head. *Asl* wh, gl, smooth→ wrinkled in 3w. a slight pinkish tinge. *P*+, 1 br. → pink, 3 w. flesh pink. *M* Co—, alk, in old cults, pinkish tinge on surface. *B* turbid△, surface memb. thick, tenacious → clear. Nitrites—. *Hab.* Sewage.

II. Non-motile.

  A. Gelatin Colonies filamentous-flocose. Pigment red-brick red.

    283. Bact. furrugineus (Dyar).

B. 0.6:1.0, pairs-chains. Growth red·brick red, crusty, gran, scarcely wrinkled; grow slowly. *G cs* (b) like a round tuft of cotton. *B* no memb. *M* Co—, reaction unchangedr brick red growth on surface. Nitrites—, *Lac. Lit* blue. *Hab.* air.

  B. Gelatin Colonies not characterized as above.

1. Pigment, bright red, carmine-blood red.
   a. Growth on fresh (acid) potato (?).
      284. Bact. finitimus ruber (Dyar.)
B. o.5:o.6-1.o, singly-chains of 3-4. *G cs* round entire, smooth. *Growth* on solid media reddish pink—bright red. *B* no memb. *Nitrites— Lac. Lit* blue. *Hab.* Air.
   b. Scarcely grow on fresh (acid) potato.
      285. Bact. Havaniensis (Sternberg).
B. Short ovals, o.4-o.5, usually in pairs, almost a caccus. 20° △. *G cs* round, small, transl, blood red. *G st* carmine layer on surface. *Asl* thick, carmine, mo, gl. *Hab.* Yellow fever cadavers?
   2. Pigment, brick red.
      286. Bact. latericius (Eisenberg).
B. 3-5x's breadth—filaments. *G cs* (2) thick, brick-red. *G st* slight growth. *P* brick red. *Hab.* Water.
   3. Pigment bright pink—salmon pink.
      287. Bact. rhodochrous (Dyar).
B. o.5-o.6-1, singly-chains. Pigment bright pink. *G cs* round, entire, smooth. *B.* no memb. *Lac. Lit* blue. *Nitrites—. Hab.* air.
      288. Bact. salmoneus (Dyar).
Identical with preceding except pigment is a salmon pink. *Hab.* Air.

CLASS XVIII. PIGMENT BROWNISH, BLACK, GRAY ON GELATIN.

1. Gelatin Liquefied.

A. Motile.
      289. Bact. cyano fuscus (Beyerinck).
B. o.2-o6 and one-half as thick. Cults. in solutions containing o.5 % peptone, green, bl— brsh black. *G cs* surrounded by a black zone in which crystals are formed. *Hab* isolated from glue, cheese.
      290. B. violaceous sacchari (Ager).
B. short, o.5: o.7-1.o, singly—short chains. Gel. liq. quickly. *M.* coagulated. *Asl* thin, transl. green, fluor. in medium. *PR* decol; *Lac. Lit* red; *Nitrites—.* A green fluor and blackish color in the presence of glucose, lactose, glycerine. A violaceous pigment especially noted in old milk cults. *Hab.* Air.

B. Non-Motile.
      291 Bact. glaucus (Maschek).
B. slender rods *G cs* round, entire, center gr, edge brown and radically folded + liq. *G st* gr bacterial masses. *A sl* and *P* gr. *Hab.* water.

II. Gelatin not Liquefied.

292 Bact. fuscus limbatus (Scheibenzucker).

B. short *G cs* brsh clumps. *G st* (1) uniform growth with short outgrowths (2) spdg $\searrow$, gel. near stab $\searrow$ brsh. Asl br color of medium, *P* brsh. *Hab* isolated from decayed eggs.

CLASS XIX. PIGMENT BLUE–VIOLET ON GELATIN OR AGAR.

III. Gelatin Not Liquefied.

A Motile.

1 On Potato pigment violet.

a On Agar violet.

293 Bact. violaceus Laurentius (Jordan-Dyar).

B. 0.7 : 3-4.6. *B* pigment formed only in presence of nitrates, nitrates reduced. *P* violet. *M*. Co., ac. (In + P R ±) Dyar. Hab. water.

294 Bact. janthinus (Zopf).

B. middle sized. *G cs* milk wh → violet, liq $\searrow$. *G st* only surface growth. *B* memb, violet. *A sl* and *P*. violet. *Hab*. water.

b On Agar blue black.

295 Bact. lividus (Plagge-Proskauer).

B. slender, medium chromogenic, blue black. *G cs* resemble drops of ink, gel. liq $\searrow$ with a blsh violet deposit *G st* (1) $\triangle$ (2) violet, liq $\triangle$. *A sl* blue black. *P* violet. *Hab*. water.

2 On Potato pigment grayish blue-dark blue.

296 Bact. coeruleus (Voges).

B. 0.8 : 1-1.4. Flag. polar, $37°+$ with pigment. *G cs* (2) typhi like → gr bl, liq $\searrow$. *G st* $\searrow$. *B* memb, gr. *M* Co +, cream sky blue. *P*. grsh bl-d bl. *Hab*. water.

3 On Potato pigment brownish.

297 Bact. amethystinus mobilis (Germano).

B. slender, long, not in threads, $37\frac{0}{-}$. *G cs* membranaceus, liq $\triangle$ → violet. *B* memb, crumpled. *A sl* $\searrow$. *P* brsh. *M*. Co +. *Hab* air.

B. Non-motile.

1. On potato dirty white to olive green.

298. Bact. amethystinus (Eisenberg).

B. 0.5:1-1.5. $37°$, *G cs* typhi like → d violet, liq $\searrow$. *Asl* uncolored → d. violet, metalic, crumpled. *P* dirty wh-olive green. *B* memb, fluid brsh. *Hab*. Water.

2. On Potato dark blue.

299. Bact. coeruleus (Smith).

B. o.5:2-2.5 chains, liq$\triangle$. *Cs* blsh.

G st funnel, in depth a few *Cs*. *Asl* blsh.

P dark blue. *Hab*. Water.

II. Gelatin not Liquefied.

A. Motile.

300· Bact. indigoferus (Voges).

B. o.6:1.8, single. Fla. polar *G cs* (1) small (2) flat blue.

G st (1) colorless (2) flat, gl. *B*$\triangle$, memb, bl. *M* Co—, surface bl gr. *Asl* d bl. *P* gsh bl. Path—.

Pigment sol. in $H_2 S O_4$=br; in HNO_3=ylsh; in HC *l*=blshl; addition of NH_4HO has no effect on pigment. *Hab*. Water.

301. Bact. indigonaceus (Schneider.)

Diff. from preceding. *P* deep indigo bl, on *B* no memb. Pigment solutions as above except that NH_4HO renders the H Cl sol. colorless. Pigment insoluable in water and alcohol. *Hab* Water.

B. Non-motile.

302. Bact. beriolinensis Indicus (Clässen).

B. like typi in size, pairs—3's, delicate capsule.

G *cs* 3d, 1 m.m., grsh wh + indigo bl, irreg. bulged.

G st (2) irreg. thin, blsh. *A sl* elev, limtd, edges deep bl. *P* elev, d bl. *Hab*. Water.

CLASS XX. PIGMENT GREENISH.

303. Bact. allii (Griffiths).

B. 2.5:5-7, singly-pairs. *A sl* thick, green layer, pigment soluble in alcohol. *Hab* isolated from decaying onions.

CLASS XXI. FLUORESCENT BACTERIA, GELATIN LIQUEFIED. MOTILE.

I. Grow on ordinary Potato.

A. Gelatin liquefied slowly.

1. Growth on potato greenish.

304. Bact. viridis (Lesage).

B. o.75: 1-2.4—threads: op. 35°. *G cs* (2) thin, spdg, erose, green fluor. *Gst* only surface growth. *P* d green, rarely reddish, odour of old urine. *Path* rabbits, inven, or by feeding " green diarrhoea." *Hab* assoc, with " green diarrhoea," of children.

2. Growth on potato brownish or becoming eventually so.

a. Grow well at 37° C.

* Very small, slender rods (one-half size of tubercle B.)
305. Bact. smaragdino foetidus (Reimann).

B. as above, somewhat bent; 20° $\triangle$, op. 37°. *Gst* liq along (1) $\triangle$. g. fluor. *A cs* irreg, fluor. *P* chocolate br. *Hab* isolated from nasal secretions in ozaena.
** Much larger rods—coccoid forms, variable.
306. Bact. proteus fluorescens (H. Jäger).

B. variable, cocci—threads, variable also in thickness. *G cs* Proteus, coli and typhi like, green fluor, cults, stink as in Proteus. *G st* often gas. *A sl* warty drops—+ thick, ylsh wh, with g. fluor. *P* slimy lyl—+ d br. *Path. mice* o.1 c.c. inper—sbc, death 3 d—2 w; fatty degeneration of liver and kidneys, spleen enlarged, intestines hemorrhagic. B in organs. *Hab.* isolated from persons in icterus with nephritis, from urinary sediment, etc.
307. Bact. No. 21 Conn.

B. 0.8: 2.0—chains. (20°—35°)+. *G cs* small, opaque liq. + —+ large gsh, gran, with a central gran, nucleus. *Gst* crateriform—broad funnel, pel, gel, clear green. *Gsl* thin, wh. transp, mo, spdg , agar—+ gsh. *P.* thin, mo, brsh. *M* Co, 20—35, alk, pep. $\triangle$. *B* turbid, pel $\searrow$. sed—+. liq. $\smile$ gsh.—+ tenacious pel. on surface. *Hab.* Milk.

b. Do not grow at 37° C.
308. Bact. fluorescens Schuylkilliensis (Wright).

B. small, short, rounded, pairs, threads. *Gcs* (2) 2 d. 1.5 m.m, gr wh, transl, (b) brsh center, border thin, radiating $\triangle$—+ g wh—bl g fluor. *Gst* liq. $\triangle$. crateriform, bl g fluor. *A sl* grsh, transl, agar fluor. *B* turbid, pel. $\triangle$, bl g. fluor. *P.* brsh, elev. spdg. *Lit. M* Co. $\searrow$, decol $\triangle$. *G lB* no gas. *In?* *Hab.* Water.

B. Gelatin liquefied quickly.
1. Gelatin stab cultures arborescent, potato cultures flocose—curled.
309. Bact. leucaemiae canis (Lucet).

B. slender, 3.0 long. *Gst* (1) lateral outgrowthst (2) liq. and fluor of surface growth. *P.* +, flocose curled. *Path rabbits* death 10 d nodular formation in inner organs containing B. *Guineas* immune. *Hab.* isolated from a dog with leucocythemia.
2 Cultures on gelatin and potato not characterized as above.
a Cultures give a decided aromatic odor.
310. Bact. chromo aromaticus (Sternberg).

B. middle sized. *G st* green, with ylsh wh memb. *A sl* thin, wh. *P* br. *Path rabbits,* inven, death 2-3w; pneumonia, pleuritis, pericarditis. *Hab* isolated from a hog with broncho pneumonia, enteritis.

b   Odor not characterized as above.

* Milk coagulated.

311   Bact. pyocyaneus (Gessard).

(B. of green or blue pus.)

B. o.4 : 1-4, variable, flag. polar. *G cs.* (1) round-oval, ylsh gsh-moru-loid. (2) round, thin, spdg, erose + liq, often with clear outer zone, often cilliate at the periphery, variable. *G sl* crateriform-succate, gyl— blg, fluor. *A sl* spdg, gl, ylg—g, fluor. *B* turbid, ylg, fluor, pel. *M* Co, pep, alk, ylg, fluor. ..⌐ ylsh, gl·br. In—, H₂S—, *G lB* no gas. (Lac Lit. bl, *PR* deepened, nitrites +) Dyar. *Path rabbits* and *guineas* △ ; large doses. sbc-inven, death 24hrs—many weeks, in small doses sbc ; supuration-hemorrhagic odema, etc.

Hab in mouth, intestines and on the skin in man, in blue pus, etc.

312   Bact. fluorescens mutabilis (Wright), *Hab.* water, also Bact. *No. 21* (Conn), *Hab.* milk, are probably only varieties of 311 chromogenic function weaker than 311. Cul-tural characters otherswise, within the range of probable normal variations, indistinguishable from the latter.

** Milk not coagulated.

313   Bact. fluorescens liquefaciens (Flügge).

B. butyri fluroescens, Lafar.

B. fluorescens nivalis, Schmolck.

B. viscosus, Frankland.

B. fluorescens minutissimus, Unna.

B. melochlorus, Winkler-Schroter.

Characters quite identical with B pyocyaneus and variable.

Hab. water, decomposing fluids, etc.

II.   Little or No Growth on Ordinary Potato (Marine Bacteria).

Phosphorescent.

A.   Do not Stain by Gram's Method.

1.   Phosphorescence bluish.

314.   Bact. phosphorescens indicus (B. Fischer).

(Photobacterium indicum Beyerinck).

B. o.6-o.8:2—bent threads. *G cs* (1) round. entire blsh-sea g. (2) gran, brsh, borders undulating, sinking slowly in liq gel. *G sl* (1) ⌐(2) napiform with air bubble.

A sl gr wh. *P.* no growth except when cooked in salt water, Phosphor-escence blsh disappearing in old cults. Grow well in sea water on fish, meat, blood, egg, with a blsh phosph. *Path* negative.

Hab isolated from phosph sea water West Indies.

315. Bact. phosphorescens indigenus (Fischer).
(Photobacterium, Fischeri, Beyerinck).

Cult. characters as before. Diff. liq gel rather more slowly, phosphorescence less intense, the latter absent when grown on flesh media as above. *Hab* Sea water.

2. Phosphorescence silvery.

316. Bact. luminosus (Beyerinck).
(Photobacterium luminosum, Beyerinck).
(B. argenteo phosphorescens liquefaciens, Kats.

B. 0.6:2.0 bent curved rods—threads, variable.

G sl liq. quickly, later generations slowly; grow best in 3.5%, N׀ Cl gel. Phosphorescence weak, silvery; does not appear on meat and *P* cults, evident in sea water on fish and in 3.5% Na Cl sugar free gel.
Hab sea water.

CLASS XXII. FLUOROESCENT BACTERIA, GELATIN LIQUEFIED, NON-MOTILE.

I. Gelatin Liquefied slowly and feebly.

317. Bact. fluorescens immobilis.
(B. fluorescens non-liquefaciens, Eisenberg).
(B. scissus, Frankland).

B 0.3-0.5:1·2—threads. *G cs* coli like, g fluor in gel, slight liq of gel. *Hab*. air, water.

II. Gelatin Liquefied quickly.

318. Bact. graveolens (Bordoni-Uffreduzzi).

B. short 0.8 *G cs* irreg. g liq quickly, bad odor.

P gr, stinking. *Hab* isolated from skin of man between the toes.

CLASS XXIII. FLURESCENT BACTERIA, GELATIN NOT LIQUEFIED. MOTILE.

I. Phosphorescent, Marine Bacteria.

319 Bact. argenteo phosphorescens (Katz).

B. 0.6-08 : 2.5 Gm +. *G cs* ylsh.

Phosphorescence silver wh with a gsh tone, not very strong. *Hab*. from sea water, Australia, also on marine animals.

II. Fluorescent, Land, Saprophytic Bacteria.

A Milk Coagulated.

320. Bact. rugosus (Wright).

B. medium, rounded, pairs-chains-threads. *G cs* 3d. 3-4m.m, transl, grsh, elev △, irreg-sinuous sharp, radiately rugose with smooth border. *G sl* grg, dense, limtd, delicately wrinkled-reticulate, gel faint g. *Ac G* +, *Asl* transl, limtd 」, gr-grsh wh, delicate wrinkles, agar → gsh. *B* turbid, pel, sed. *P* mo, gl, br. *Lit M* Co +, ac, *G lB* no gas. *In* △, nitrites—, (35°-36°)—. *Hab.* water.

B Milk not coagulated.

1 Milk rendered slightly acid or litmus milk becomes slightly pinkish.

321 Bact. fluorescens incognitus (Wright).

B. medium, short, rounded, pairs-threads-chains. *G cs* (1) round-oval, yl br, gran. (2) 2-3d, thin, transl, ± 3m.m., edge irreg-wavy, coli like, (b) △ gran, △ yl br. center with yl br. nucleus, marmorated, older cs. 6-8m.m, gsh tint, gel. acquires a blg. fluor. *G sl* thin, transl, △ gsh, limtd.

A sl thin, mo, transl acquires a gsh color.

B turbid, pel 」 sed → gsh. *P* mo. gl, spdg, br.

Lit M. decol 」 (after a month or so) reaction acid. *Hab.* water.

322 Bact. fluorescens foliaceus (Wright).

Dr. Wright doubtless has good grounds for making this and the preceding distinct species, a careful study of the descriptions, however, do not reveal upon what specific differentiation can be well based. The features of distinction appear to be *G sl* a central furrow with also laterals, giving a leaf like etching of the medium. Surface *G cs* with heavy br. radial stripes. *Lit M.* deeper bl, alk → acid, pink. *In* *Hab.* water.

323 Bact. putidum see No. 314. Var B. virescens (Frick-Dyar).

2. Milk rendered alkaline in reaction.

a. Gelatin surface colonies thin, flat.

* Do not grow at body temperatures. Milk reduced, blue.

324. Bact. cyanogenus (Ehrenberg).

B. syncyanum, Schröter.

B. of blue milk.

B. 0.3-05:1-4. Fla. polar. *G cs* (1) small, round 0.8-1.1:1.6-5, dark (2) large thin spdg 0.8-1.3, with erose edges; coli type, rarely of aerogenes types.

G sl (1) little growth. (2) wh—blsh gr.

P slimy, bl gr-br. *A sl* gr wh, agar variously colored, *B* turbid, gr g +
bl g, pel±

G lB no gas, ac⟨. *In* tr. H₂S—, NH₃+. *M* alk. blue.

Hab. Blue milk.

** Grow at body temperatures.

† Gelatin surface colonies become fillamentous on their
borders.

325. Bact. striatus viridis (Ravenel).

B. slender, variable lengths, rods irreg, stained like diptheria, singly-
pairs; Fla. polar. *G cs* (1) round, ylsh, gran, entire→ brsh (2) 5d.

1 m.m. zoned, central nucleus+zone of granules+outer filamentous
border. *A sl* very thin, agar→ g-ylsh g.

G st (1) ⟨(2) wh. 2 m.m. in 7d, elev.

P mo. gl→ chocolate br. *B* turbid. →7d. g. *PR* ±.

Lit M 1od. more bl. in 5 W. decol. *G l* B no gas. *In*—. *Hab* soil.

†† Gelatin surface colonies, sharp, entire.

326. Bact. fluorescens ovalis (Ravenel).

B. short rounded 2-3x's breadth, singly, mot. fla, polar.

G cs (1) pale gr, ⟨gran, entire (2) round, 1 gr, ⟩gran. entire → 3d,
1 m.m, bl wh-ylsh g.

G st (2) (1) filiform (2) wh. button with irreg. leafy margin.

A sl thin, g sh wh, limtd, agar → faint g.

P thin, mo, honey y→ ylsh br. B. turbid, flaky, pel on surface, gsh tint
to B. *Lit* M deeper bl, alk.

G l B. no gas. In—, *Hab* soil.

327. B. viridescens non-liquefaciens (Ravenel).

There seems to be no essential difference from descriptions between this
form and the preceding.

b. Gelatin surface colonies convex.

328. Bact. fluorescens convexus (Wright.)

B. medium, short thick, rounded. *Gcs* (1) round, sharp, gran   (2)
round, convex, gl, lgsh transl, 2-3 m.m.; gel. acquires a blg fluor.
Gst (2) elev, gl, lg, gel→ blg fluor. *A cG* +, fluor—, *A sl* transl, mo,
gl, lgsh, agar→ gsh. *B.* turbid,. gsh. *P.* pale br, spdg. *Lit M* Co—,
alk, color deepened. *GlB* no gas. *In ?* 35°—36°, little or no growth.
Hab. Water.

3 Milk reaction not changed.

a Reddish granules in the rods.

329 Bact. erythrosporus (F. Cohn).

B. slender threads, at 20° in every rod 2-8 reddish ovoid granules. *G cs*

(1) lobed, undulately channeled, g fluor. (2) round, irreg, brsh, with a weak radial stripening. *G st* (1) equal growth (2) flat. *P* spdg. ⅃, rdsh + nut br. *Hab* flesh infusion, water.

b Rods not characterized as above.

330 Bact. putidum (Flügge).

B. fluorescens putidus, Flügge.
B. fluorescens non-liquefaciens.
B. fluorescens longus, Zim.
B. fluorescens tenuis, Zim.
B. viridis pallescens, Frick.
B. virescens, Frick.
B. viridis, Lesage? No. 304.
B. aquatilis fluorescens, Lustig.

B 0.4-0.8 : 1.6-5. Fla polar, Gm—, ae. *G cs* (1) round, oval, ylsh, homo. (2) (a) 2d. 2-3m.m.transl, erose, gl, ylsh g, gel ylg fluor + 10 m.m. (b) of coli type. *G st* (1) filiform (2) thin, wh, ylg, gel fluor ylg. *M* Co—, am, gradually dssolved with ylg color. *Hab*, water, air, etc.

CLASS XPIV. FLUORESCENT BACTERIA, GELATIN, NOT LIQUEFIED, NON-MOTILE.

I. Do no Grow on ordinary (acid) Potato.

A Grow in milk

331 Bact. smaragdino phosphorescens (Katz).

B. 1.0 : 2.0, somewhat pointed ends, solitary—pairs—coccoid; ae, *G cs* (1) 18h oval, entire, o.15 m.m., broad center + narrow zone + outer marginal zone. (2) 18h. thin, 1 gr—ylsh, gran ⅃, margin transl, dentate ⅃; 2od. 2 m.m. flat irreg, center ylsh with slate colored zone. *st* (1) thin (2) flat, stearine lustre + 5 m.m. *A sl* ⅃. *B* no growth without Na Cl. *M* gl. sticky layer on surface. *P* no growth when acid, neutralized with sodic phosphate=thin, brsh. Cults. added to sea water caused a decided phosphorescence. *Hab*. isolated from herring, Sydney.

B No growth in milk.

332 Bact. phosphorescens (Fisher).

(Photobacterium phosphorescens, Beyerinck.)

According to Kuse. See Flügge M. 332, ident. with above. *Hab*. isolated from fish, etc.

323. Bact. phosphorescens Giardi (Giard-Billet).

B. and cults. like preceeding but—smaller and more coccoid. *Hab.* isolated from crustaceae.

II. Grow on ordinary acid potato.

A. Grow in nutrient bouillon in absence of NaCl.

1. Bouillon fluorescent, potato becomes black.

334. Bact. fluorescens of Lepierre.

B. 0.5: 2-3. Gm—. *Gcs* round—ylsh br.—+ 3d fluor, in 5 d. Cs. are green. *P* no fluor—+ black. *B.* fluor. of medium. *M* Co—, alk, no fluor. *In*—. *Gl B* no gas. Op. 20°—30°. No fluor. at 37°. *Path.* guineas death 1-6 d; liver abscess, peritonitis. *Hab.* Water.

2. Bouillon not colored. Potato light brown. Colonies of aerogenes type.

335. Bact. fluorescens (Crassus).

(B. Iris. Frick).

B. very small slender. *Gcs* of aerogenes type. *Gst* (2) an aerogene like bead, g fluor—ylsh br— dg and fluor. *P l* br. *B* no memb. not colored. *Hab.* air, water.

B. Do not grow in nutrient bouillon in abscence of NaCl.

336. Bact; phosphorescens Pflügeri.

(Photob. Pflügeri, Beyerinck) (B. phosph. gelidus, Forster).

Morph. and in cults. like *B. phosph.* Fischer but somewhat longer and more slender. Grows on potato. *Hab.* isolated from fish, etc.

CLASS XXV. ANAEROBIC (OBLIGATIVE) BACTERIUM.

I. Gelatin liquefied.

337. Bact. anaerobic No. II (Flügge).

B. rather thick. *Gcs* not like, liq. rapid. *G cs* bryl, compact, with outgrowths \, much gas, rancid small. *M* Co. without gas. *Hab.* Milk.

II. Gelatin not liquefied.

A. Grow at room temperatures and in nutrient gelatin.

1. Gas produced in ordinary nutrient gelatin.

a. Bacilli surrounded by a capsule.

338. Bact. aerogenes capsulatus (Welch).

B. somewhat thicker than anthrax, chains, with a capsule. *Gel.* cults give much gas. *G cs* 1-2 m.m. gr wh, oval—irreg with a few outgrowths. *M* Co. faint odour of old cheese. *P* gr wh. *Path.* negative. *Hab.* isolated from blood in a case of anurism of the aorta.

74

b. Bacilli not surrounded by a capsule.
339. Bact. I (Sanfelice).
B. of variable length, with bladery swellings. G cs round, gl. finely gran, edge flocose—B. zopfi like outgrowths. G cs thick, flocose—tomentose. G st beaded, gas. Hab. isolated from putrifying flesh.
2. Gas not produced in ordinary nutrient gelatin.
a. Large thick rods, over 1.0 microns.
340. Bact. emphysematosus.
(B. of Gasphlegmone, E. Fränkel).
B. somewhat thicker than anthrax, threads. Gm G, with addition of Gl yc. and formate of soda, active growth, with much bad smelling gas. G l B much gas. B turb.
Path guineas agar cult in aq. sbc. local non-supurating inflamation —+ necrosis, sometimes a spreading to peritoneum and pleura. Hab isolated from a gaseous plegmon, etc.
b. Very slender rods, 0.3 microns.
341. Bact. filiformis Havaniensis (Sternberg).
B. long, slender, o.3 diam. G st (1)△(2) no growth on surface. A sl , milk wh. branched. P—, Path (rabbits, guineas)—. Hab. liver yellow fever cadaver.
B. No growth at room temperature or in nutrient gelatin.
1. Motile.
342. Bact. oedematis thermophilus (Novy.)
B. 0.8-0.9:2.5-5, Fla. Gm , G cs · flocose.
Sugar B gas, without smell or that of butyric acid, reduces litmus. No growth below 24° C.
Path very path. for mice, rats, guineas, rabbits, cats, pigeons by sbc. injec. of o.1-0.25 c.c.; odema containing much gas, abdominal muscles red with hemorrhagic flecs; plural and abdominal cavities contain colorless exudate.
Hab. isolated from a guinea inoc. with a contaminated nuclein solution.
2. Non-motile.
343. Bact. cadaveris (Sternberg).
B. 1.2:1.5·4 short threads. Do not grow in gel.
Glyc B, ac, no gas.
Path guineas death with odema by inoc. with a section of liver containing the B. Cultures but slightly path.
343 a Bact. pyogenes anaerobius (Fuchs).
B. large. Do not grow below 22°; causes a stinking supuration in rabbits.

Hab. isolated from stinking pus from a rabbit which died spontaneously.

344   Bact. pyogenes anacrobius (Fuchs).

B. large.   Do not grow below 22°; causes a stinking supuration in rabbits.

Hab. isolated from stinking pus from a rabbit which died spontaneously.

APPENDIX.

Bacteria not sufficiently described or of such habit as to make their classification by the foregoing scheme impossible.

I.   Bacteria Associated with Diseases of Plants.

345   Bact. amylovorus (Burrill)

B.   0.5-0.7 : 1.25.   On plant decoctions zoogloœ, liq—, produces Co_2, butyric acid and alcohol.   *Hab* associated with "pear blight."

346   Bact hyacinthi (Wakker).

Cult. characters not known.   *Hab* assoc with disease of hyacinths, decay of bulbs and yellow lines and blotches on the leaves; morbid tissue contain numerous bacteria.

347   Bact. hyacinthi septicus (Heinz).

B.   1 : 5.   Mo +, *G cs* (1) spdg. transl. (2) oval, compact, ylsh.   *G st* nail flat top.   *G sl* transl.   *P* dark yl, slimy.   *Path.* infection expts +.   *Hab.* decayed hyacinth bulbs.

348   Bact. mori (Boyer-Lambert).

Cultural characters not given.   *Hab* disease of the mulberry, brown spots on leaves and branches containing B.   *Path* inoc. expts. positive.

349   Bact. oleae (Trevisan).

Gultural characters not given.   *Hab* assoc. with olive galls, tuberculosis of the olive; inoc. expts. positive.

350   Bact. pini (Vuillemin).

B.   0.6-0.8 : 1.5-2.5 in zoogloea.   Stain with difficulty inoc. expts. negative.   *Hab* assoc. with galls on Alpine pine.

351   Bact. sorghi (Burrill).

B.   0.8-1.0 : 1.5-3-chains *A sl* and *P* mother of pearl.   *Hab* assoc. with disease of sorghum.

352   Bact. uvae (Cugini-Macchiati).

B.   0.2.5:3.4, Mot.; liq +.   *P* honey yl cs.   *Hab* assoc. with a disease of the grape; berries become brown, dry and fall off.

353   Bact. zea (Burrill).

(B. secalis.)

B.   0.65 : 0.8-1.6 gel. liq.   *A sl* opaque gl.   *Hab* assoc. with a disease of Indian corn.   Inoc. expts. negative.

354 Bact. betae (Arthur-Golden).

B. 2x's breadth. mo+, *gel.* whsh → lyl → liq _, alk. *A sl* lylsh *Pasteur's solution* turbid, gr. sed. Inoc. expts. positive in part. *Hub* assoc. with a disease of sugar beets.

INDEX AND MOST IMPORTANT LITERATURE OF SPECIES OF GENUS BACTERIUM.

*Number of species in the text.

345 Amylovorus (Burrill), Arthur, N. Y. Expt. Sta., 1884, 357; 1886, 259.

Waite, Bot. Gaz. XVI, (1891), 259.

337 Anaerobic No. 2, Flügge. Z. H. XVII, 2.

223 Annulatus (Wright), M. N. A. S. VII, 443, 1895.

153 Antenniformis (Ravenel), M. N. A. S. VIII, 25, 1896.

121 Apthosus (Siegel), D. M. W., No. 49-94, 1891.

182 Aquatilis (Frankland), Z. H. VI.

154 Aquatilis communis (Zim), M.

330 *Aquatilis fluorescens* (Lustig), M.

201 Aquatilis radiatus (Zim) M.

62 Aquatilis solidus (Lustig) M.

30 Aquatilis sulcatus quartus (Weichselbaum) Osterreichisches, Sanitätwesen, 1889.

Dyar N. Y. Acad.Sci. VIII, 359, 1895.

42 Aquatalis sulcatus (Weichselbaum) Osterreichisches Sanitätwesen, 1889.

Del Rio. A. H. XXII, 2; Vaughan-Sternberg M. Nos. 477, 480, 483.

Bockmann A. Expt. No. 33.

Nicolle-Refik. A. P. 1896, 242.

233 Arborescens (Frankland) Z. H. VI.

218 Arborescens non-liquefaciens (Ravenel) M. N. A. S. VIII, 39, 1896.

319 Argento phosphorescens (Katz) C. IX.

199 Aromaticus (Pammel) Iowa Expt. Sta. B. 21, 1893, p 792.

252 Aurantiacus (Frankland) Z. H. VI.

251 Aureo flavus (Adametz) M.

225 Aurescens (Ravenel) M. N. A. S. VIII, 8, 1896.

251 *Au eus* (Adametz).

85 Avicidum (Kitt).

134 *Bact. of Babes and Oprescu.*

302 Berolinensis indicus (Clässen) C. VII, 13.

354 Betae (Arthur-Golden).

77 *Bipolare mullocidum*, Kitt.

89 Bovisepticus, Kitt. Sitzber. d. Ges. f. morph., Münch, 1885.

Hueppe B. K. W., 1886, 44.

78

Oreste-Armanni, B. J., 1887, 208; Poels, B. J., 1887, 209.

Bunzl-Federn C. IX, 24; Jensen, Monatsch, prakt. Thierheilk II, B. J. 1890, 188.
18 Breslaviensis (Van Ermenghem) Trav. Lab. d'Hyg. de Grand Bruxelles. 1892, Bd. 1.3, Kaensche Z. II. XXII, 1896, 53. Holst, C. XVII, 717, 1895.
266 Brunnes flavus (Dyar) N. Y. Acad. Sci. VIII, 362.
155–203 Buccalis fortuitus (Vignal) Archiv. Phys. VIII, 342.
245 Buccalis minutus (Vignal) Archiv. Phys. VIII, 1886; Sternberg, M. 1892, 643.
89 *Büffelseuche* (Oreste-Armani).
106 Buffalo plague (Ratz) C. XX, 290, 1896.
313 *Butyri fluorescens* (Lafar) A. II., XIII.
343 Cadaveris (Sternberg), M. 1892.
253 Campestris (Pammel), C. C. I, 17.
82 *Canalis capsulatus* (Mori), Z. II. IV, 47.
115 *Canalis parvus* (Mori).
51 Canary bird septicaemia (Rieck), Z. T., 1889.
73 *Candicans* (Frankland) Z. VI.
82 *Capsule Bacillus* of Nicolaier C. XVI, 601.
82 " R. Pfeiffer, Z. II. VI, 115, 1889.
82 " Mandry, C. VI, 570.
82 " Kockel, F. M. IX, 1896.
82 " Dinugesen, C. XIV.
82 " Marchand C. XV, 428.
82 " Loeb, C. X, 369.
82 *Capsulatus* (Banti) Sperimentale, 1888.
80 *Capsulatus mucosus* (Fasching).
82 *Capsulatus septicus* (Bordoni-Uffreduzi.)
272 Carneus (Tils), Z. H. VI, 282. 1890.
Caucasicus, C. III, 135.
240 Caudatus (Wright), M. N. A. S. VII, 444, 1895.
76 *Castellum*.
45 Caviae fortuitus (Sternberg) M. 1892, 650.
77 *Cavicida* (Brieger), Zeitsch. phys. Chem. VIII.
46 Cavicida Havaniensis (Sternberg), M. 1892, 425.
175 Centrifugans (Wright), M. N. A. S VII, 462, 1895.

244 *Cereus citreus* (Dor).

110 Cholera columbarum (Leclancher), A. P. 490, 1894.

85 *Cholera anatum.*

85 Cholera gallinarum.

Perroncito A. T. 1879; Pasteur. C. R. XC, 239, 1880, XCI, 673; Gaffky, M. K. G. I. 94, 1881, C. I. 305, 1887; Kitt C. I No. 10, 1887; Bunzl-Federn, C. IX 24; Karlinski, C. VII, 11.

Choleroides (Bujwid), C. XIII, 120, 1893.

13 Chologenes (Stern), D. M. W., 1893, No. 26.

310 Chromo aromaticus (Sternberg), M.; V Galtier. C. R. 106.

160 Circulans (Jordan). *loc. cit.*

222 Citrens (Ulna-Tommasoli), Manatasch, f. prakt. Dermat IX.

244 Citreus cadaveris (Strassmann-Strecker), Zeitch, f. Med'beamte, 1888.

146 Cloacae (Jordan), Mass. B. Health, 1890, 836.

147 Coadunatus (Wright), M. N. A. S., VII. 460, 1895.

299 Coeruleus (Smith), C. III, 401.

296 Coerulens (Voges), C. XIV, 303.

169 Cohaereus (Wright), M. N. A. S. VII, 464, 1895.

87 Coli anaerogenes (Lembke), A. H. XXVII, 384, C. XXI, 281, 1896.

11 Coli anidolicum (Lembke), A. H. XXVII, 384, C. XXI, 281, 1896.

133 Coli colorabilis (Naunyn), D. M. W., 1891, No. 5.

10 Coli communis (Escherich), Darmbak des Säuglings, Stuttgart, 1886; Emmerich D. M. W., No. 5; Weisser Z. H. I. (1886), 315: Lesage S. B., 1892; Snoeck Inaug Diss, Utrecht, 1892; Femlin A. H. IX (1893), 295; Kiessling Review of Lit. of, Hy. R., 1893, No. 16; Etienne S. B., 1894, 14; Taval-Lanz, C. XIV, 705; Stoecklin C. XVI, 130; Villinger, A. H. XXI, 101; Dreyfus, Inaug, Diss, Strassburg, 1894; Billings-Peckham Science, Feb., 1895; Ehrenfest A H., XXVI; Refik A. P., 1896, 242; Klie, C. XX, 49; Silvestrini, r. B. J., 1891, 249; Achard-Renault S. B., 1892, 22, r. B. J., 1892, 279; Gilbert, S. M. 1895, C. XVII 480; Muller r. C. XVII, 682; Elsner Z. H. XXI, 1895; Piorkowski C. XIX, 686, Kashida, C. XXI, 802; Dunbar Z. H. XII, 1892; Germano-Maurea, Zi. XII, 494, C. XV, 62; Ligniers r. B. J., 1894, 321; Nicolle A. P., 1895,

No. 1; Frendenreich, C. XVIII, 102; Smith A. J. M., CX,
No. 3, r. C. XVIII, 589; Abba C.XIX; Capaldi-Proskauer,
Z. H. XXIII, 452.

77 Coli immobilis.
Germano-Maurea, Zi. 12, 498, Brieger, Z. f. phys, Ch. VIII.
Lewandswsky, D. M. W., XC, 51.
17 Coli mobilis (Messea) Riv d'igiene, Roma, 1890.
77 Coli similis (Sternberg), M.
5 Conjunctivitis (Morax), A. P., 1896, 337; C. 21, p 1.
8 Conjunctivitis (Koch-Kartulins), Koch, A. K. G. III; Kartulins,
C. I, 289.
Weeks. Archiv. f. augenheilk, Bd. XVII.

127 Conn. Bact. No. 16. Storrs Expt. Sta. Rep. 1893, 51.
307 " " " 21. " " " " " 52.
131 " " " 22. " " " " " 53.
130 " " " 26. " " " " " 54.
50 " " " 27. " " " " " 54.
129 " " " 41. " " " " 1894 57.
158 " " " 46. " " " " " 80.
126a " " " 53. " " " " " 82.
125 " " " 55 " " " " " 83.
124 " " " 56. " " " " " 83.
265 Constricus (Zimmermann), M.
204 Convolutus (Wright) M. N. A. S. VII, 1895, 461.
117 Coprogenes parvus (Bienstock), Z. K. M., VIII.
26 Corn stalk disease (Billings) B. J., 1889, 184; 1891, 200.
Cubonianus (Macchiati), Malpighia, V, 1892.
Zeitsch. f. Pflanzenk, II, 43.
112 Cuniculi pneumonicus (Beck), Z. H. XV, 1893, 363.
85 Cuniculicida (Flügge).
9 Cuniculicida mobilis (Ebert-Mandry), V. A. 121.
133 Cuniculicida Havaniensis (Sternberg).
107 Cuniculicida immobilis (Smith), Jour. Comp. Med. and Surgery,
VIII, 1887, 24.
58 Cuniculicida septicus (Lucet) A. P., 1892.
105 Cuniculicida thermophilus (Lucet), A. P., 1889, 401.
226 Cuticularis (Tils), Z. H., IX, 282, 1890.
289 Cyanofuscus (Beyerinck), Bot. Zeit. XLIX, Nos. 43-47, 1891.

324 Cyanogenus (Ehrenberg).

Hueppe, M. G. K. II, 355; Heim A. K. G. V, 518.

Neelsen, B. B. III, 2. Scholl F. M., 1889. Gessard A. P., 1891, 12.

324 Cyncyanum (Schröter).

235 Decidiosus (Wright), M. N. A. S. VII, 443, 1895.

96 Decolorans major and minor (Dyar), N. Y. A. Sci., 362.

186 Delabens (Wright), M. N. A. S., VII, 456, 1895.

278 Delta (Dyar) N. Y. Acad. Sci., 368.

159 Delicatulus (Jordan) Mass. B. Health, 1890, 837.

215 Dendriticus (Lustig), M.

65 Denitrificans I. (Stutzer-Burri), C. C. I. (1895), 350.

66 Denitrificans II. (Stutzer-Burri), C. C. I. (1895), 350.

64 Denitrificans agilis (Ampola-Garino), C. C. II, 670.

181 Deovorans (Zimmerman), M.

229 Dianthi (Arthur-Bolley), Ind. Expt. Sta. Bull. 59, 1896, 22.

184 Diffusus (Frankland), Z. H. VI.

52 Diptheriae avium (Loir-Duclaux), A. P. 1894, No. 8.

109 Diptheriae columbarum (Löffler), M. K. G. II.
       Babes-Puscariu, Z. H. VIII, 376, 1890.

111 Diptheriae cuniculi (Ribbert), D. M. W., 1887, No. 8.

256 Domesticus (Dyar), N. Y. Acad. Sci., 358.

248 Dormitator (Wright), M. N. A. S. VII, 412.

163 Dubius (Bleisch).

113 Dubius pneumoniae (Bunzl-Federn) A. H. XIX, 326, 1893.

150 Duplicatus (Wright), M. N. A. S. VII., 457, 1895.

211 Dysenteriae liquefaciens (Ogata), C. XI., 624; Kamen C. XVIII., No. 14-15.

122 Dysenteriae vitulorum (Jensen), B. J., 1892, 308; C. XVIII, 653.

340 Emphysematosus (E. Fränkel), C. XIII, 1. Rosenbach (Mikroorg. d. Wundinfekt, 1884), E. Levy, Z. Ch. 32.

136 Endocarditis griseus (Weichselbaum), Zi. IV. (Zieglers Beiträge)

138 Endometritidis (Kaufmann), Germano-Maurea Zi. XII.
       Emanuel-Wittkowsky Z Gy. 32.

12 Enteritidis (Gartner), Korrespond d. allg. arztl. von Thuring, 1883.
       Karlinski C. VI, Lubarsch, V.A. 123.

276 Epsilon (Dyar), N. Y. Acad. Sci., 369.

23 Eqni intestinalis (Dyar-Keith), C. XVI, 838.

189 Maddoxi (Miquel) Ann. d. Microg., 1889–92.

192 Mallei (Loffler-Schutz), D. M. W., 1882, No. 52. Löffler, A. K.
G., I, 1886, 141. For bibliography up to 1892, see Sternberg,
M., 1892, 820; Babes, B. J., 1893, 261; Kutscher, Z. II., XXI,
156, 1895.

9 Marsiliensis (Rietsch-Jobert), C. R., C.VI, Caneva C.IX, 560.
Bunzel-Federn, C. IV. See also synonyms.

103 Martinezii (Sternberg), M.; Dyar, l. c., 364.

132 Mauseseuche, Laser.

22 Mäusetyphus.

313 Melochlorus (Winkler-Schröter).

298 Membranaceus amethystinus (Jolles), Eisenberg, M.
Membranaceus amethystinus mobilis (Germano), C.XII, 516, 1892.

24 Meningitidis (Neumann-Schäffer), V. A., C.IX.

180 Meningitidis aerogenes (Centanni), Archiv. Sci. Mediche; XVII,
1893, 1.

274 Milk red B. of.

324 Milk blue, B. of.

206 Mirabilis (Proteus), Hauser. (See Proteus vulgaris.)

57aMonachae (Tubeuf), Forst. Naturwiss, Zeitsch. I, 34, 1892, C.XII,
268, 1892.

17 Monadiformis (Messea), Rio d' igiene Roma., 1890.

21 Morbificans bovis (Basenau), A. H., XX, Ostertag, Fleischbeschau.
1892, Cotta, Hy. R., 1891, 716.

318 Mori (Boyer-Lambert), C. R., CXVII, 1893.

80 Mucosus ozonae (Abel).

220 Multipediculatus (Flügge), M.

148 Multistriatus (Wright), l. c., 462.

132 Muripestifer (Laser), C. XI, 184, XIII, 643, XV. 33.

191 Murisepticus (Flügge), Koch Wundinfectionsk. 1878. Löffler M.
K. G. I, 178, bibliography of subject.

213 Murisepticus pleomorphus (Karlinski), C. V, 6.

9 Mustelae septicus (Ebert-Schimmelbusch), V. A. CXV.

10 Neapolitanus (Emmerich).

167 Nebulosus (Wright), l. c.

41 Nexibilis (Wright), l. c.

170 Nitrificans (Burri-Stutzer), C. C. I, 735.

200 Nubilus (Frankland), Z. H. VI.

221 Ochraceus (Zimmerman), M.

56 Oedematis aerobius (Sanfelice).

342 Oedematis thermophilus (Novy), Z. H, XVII, 2.
Okada, C. IX, 442.

196 Orchiticus (Kutscher), Z. H. XXI, 158, 1895-96.

349 Oleae (Trevisan), Prillieux, Rev. gen. botanique, 1889, r. Bot.
Gazette, 1889, 208. Pierce, Jour, Mycol. VI, 1891, 148,
Zeitsch. f Pflanzenk, I, 161.

261 Ovalis (Wright), l.c. 435.

82 *Ovatus minutissimus* (Unna).
Tommasoli–Monatsch f. prakt. Dermatol, IX, 59.

234 Oxylaticus (Dyar), l.c., 369.

82 *Oxytocus perniciosus* (Flügge).

80 Ozoaena (Fusching-Abel), Abel. XIII, 161, 1893; Fasching, Sitz-
gsber. Wien, Akad. 100.
Loewenberg A. P., 1894, 292.
Abel. Z. H. XXI, 89, 1896. Paltauf, W. K. W., 1891,
52-53.
Wilde W. K. W, 1892, Nos. 1-2.
Thost D. M. W., 1886, 10; Dittrich C. II, 88.
Hayek, B. K. W., 1888, 659.

76 *Pallens.*

76 Pallescens (Henrici), Bak. des Käses, Baseler, Phil. Diss., 1894.

224 Pallulans (Wright), l.c., 445.

207 Parsini No. 7 Bact., V. A., 122.

212 Pansini No. 9 Bact. V. A., 122.

25 Paradoxus (Kruse-Pasquale), Z. H. XVI (1894), 1.

69 Pasteurianum (Hansen), C. R. Carlsb. Lab. Copenhagen 79.
Lafar. C. C. I, 129, 1895.

179 Pestifer (Frankland), Z. H. VI.

91 Pestis bubonicae (Kitasato-Yersin), A. P., 1894, 662.
Aoyama C. XIX, 1896, 481.
Yersin-Calmetti-Borrel, A. P., 1895, 589. C.
XXI, 497.

332 Phosphorescens (Fisher), Z. II, 92.

336 *Phosphorescens gelidus* (Forster).

333 "   Giardi, Giard-Billet, S. B., 89–90.

336 "   Pflugeri, Forster, C. II, 12 ; Beyerinck, C. VII, 338.

33 Pinnatus (Ravenel), l. c., 32.

156 *Piscicidus agilis* (Sieber), C. XVII, Nos. 24–25,

243 Plicatus (Zimmerman), M.

48 Phasiani septicus (E. Klein), J. P., 1893.

314 Phosphorescens indicus (Fischer), Z. H., II, 54.

315 "   indigenus (Fischer), l. c.

350 Pini (Vuillemin), C. R., C. VII, 1888.   Prillieux, Rev. gen. bot.,
          1889.

53 Pneumonia in Turkeys (MacFadyean), Ann. Inst. Nat. Agron, 1890.

82 Pneumoniae (Weichselbaum), Wien. med. Jahrb, 1886.

          Friedlander, F. M., 1883, 22 ; Pfeiffer, Z. H., VI,
          145, 1889.

          Loeb. C. X, 367, 1891 ; Grimbert, A. P., 840, 1895.

          Brieger, Zeitsch, Phy. Chem., 1883 ; Fermi, A. H.,
          X, 1890, 1.

82 *Pneumococcus of Friedlander.*

179 Pneumonicus agilis (Flügge), Schou. F. M., 1885, 15 ; Neumann,
          Z. K. M., XIII.

197 Pneumonicus liquefaciens (Arloing), C. R., 99, 109, 116.   Bull. Soc.
          Med. Vet. 48, (1894, 283, 302, 505.

57 Pneumosepticus (E. Klein), C. V., 625.

197 *Pneumobacillus liquefaciens bovis* (Arloing).

31 Primus Fullesi, (Dyar), l c., 360.

271 Prodigiosus(Flügge),Schottelius, C. II, 439; Wasserzug. A. P., 1888.

          Kübler, C. V, 383 ; Scheurlen, A. H., XXVI.

          Bordoni-Uffreduzzi, Hy. R., 1894 ; Gorini, Hy. R.
          1893, 381.

          Fermi, A. H., X ; Erdmann Jour. prakt. Chem. 66

          Schroter B. J., I, 2 ; Griffiths, C. R., 92.

306 Proteus fluorescens (H. Jäger), Z. H., VII, 3 ; Nonwerck, M. M.
          W., 1888, 35.

          v. Stirl, D. M. W., 1889, 39 ; Ducamp (Revue
          Méd., 90.6.

          Hüeber, ibid., 1888–90.   Globig. ibid. 1891.

179 *Schou.* Bact. of F. M., 1885, No. 15.

19 Schweinpest (Bang-Selander).

190 *Schweinrotlauf,* B.

88 *Schweineseuche Deutchen.*

317 *Scissus* (Frankland), Z. II. VI.

253 *Secales,* Syn B. zea.

99 Secundus Fullesi, Dyar, l. c., 359.

209 Septicus (Proteus) (Babes), Septische Prozesse des Kindesalters, Leipzig, 1889.

3 Septicus accuminatus (Sternberg), M. 1892, 472, Babes l. c.

115   "   agrigenus (Flugge), M. 257. Okada, C.IX, 442.

108   "   *hominis* (Mironoff).

161   "   putidus (Roger), Revue d. Med. 1893, 103, 693. C. R
            Soc. Biol, 1893 ; 707, r. B. J., 1893, 339.

16 Sinuosus (Wright), l. c., 440.

305 Smaragdino foetidus (Reimann), Würzburg, Diss. 1887.

331   "   phosphorescens (Katz), C. IX.

40 Solanacearum (Smith).

28 Solitarius (Ravenel), l. c., 29.

78 Sordidus (Dyar), l. c., 379.

351 Sorghi (Burrill), Kellermann, Rep. Kas. Expt. Sta., 1889.

254 Spiniferus (Tommasoli), Mon. Prakt. Derm., IX.

92 Sporadic pneumonia in cattle (Smith), U. S. Dept. Ag. Bureau
   Animal Industry, 1895, 136.

140 Sputigenes crassus (Kreibohm), Flugge M. ; Babes. C.VII, 600.
    Dyar, l. c., 356.

143 Sputigenes tenuis (Pausini), V. A., 122.

152 Stoloniferus (Pohl), C.XI, 142.

263 Striatus flavus (r. Besser), Zi. B. VI.

325   "   viridis (Ravenel), l. c., 22.

63 Stolanatus (Adametz), M.

249 Subflavus (Zimmermann), M.

255 Subochraceus (Dyar), l. c., 358.

19 Suipestifer (Salmon-Smith), Rept. Bur. Ann. Industry, U. S.
   Dept. Ag. 1885–91 ; Smith, C. IX, 252, 339 ; Z. H. X., 480.
   Selander, C. III, 360.

88 Suisepticus (Schütz), Loffler-Schütz, A. K. G. I.
   Cornil-Chantemesse, Bull. Méd., 1887, 85.

Smith, C. X. Bleisch- Fiedeler, Z. H. VI.

Bunzl-Federn, C. IX, No. 24.

185 Sulcatus liquefaciens (Kruse), Flügge, Die Microrganismen, 1896, 318.

208 Sulfureus (Proteus) (Holschewnikoff), F. M., 1889.

174 Superficialis (Jordan), Mass. B. Health, 1890, 833.

9 *Swine plague* (Billings).

88 *Swine plague*, Salmon-Smith.

162 Tachytonum (Fischer), D. M. W., 1891, 26-28 r. B. J. 1894, 437.

131 Theta (Dyar), l. c., 375.

69 *Tholoideum* (Gessner), A. H. IX.

98 Tiogensis (Wright), l. c., 441.

32 Tracheiphilus (Smith), C. C. I, 364.

183 Trambustii (Transbusti-Galeotti), C. XI, 717, 1892.

227 Tremelloides (Tils), Z. H. IX, 282, 1890.

Truthahn pneumonia (MacFadyean), B. J., 1893, 147.

36 Typhi (Eberth-Goffky), Eberth, V. A. 1881-83; Koch. M. K. G. I. 15. Goffky, M. K. G. II. Bibliography of Species, A. K. G. XI, 207.

22 Typhi murium (Löffler), C. XI, 129; XII, I; Lunkewitsch, C.XV, 845.

Kornanth C.XVI, 104; Mereschowsky C. XVI, 612, XVII, 742.

73 *Ubiquitus* (Jordan), Mass. B. Health, 1890.

7 Ulceris cancrosi (Ducrey), Mon. f. Dermat, 1889, heft IX, r. B J., 1889, 238; C. XVIII, 290 Petersen C. XIII; Unna C. XVIII, 234, Mon. f. Derm, '92.

141 Ureae (Leube), V. A., 100.

84 Ureae (Jaksch), Dyar, l. c., 357; Jaksch. Zeitsch. f. phys. Chem. V. 395; Luebe-Grassar, V. A. C. 556.

352 Uvae (Cugini-Macchiati), r. Zeitsch. f, Pfanzenk, 1891, 1894.

94 Vacuolatus (Dyar), l. c., 357.

6 Vaginae (Döderlein), Das Scheidensecret. u. seine Bedeutung. f. d, Purerperalfieber, Leipzig, 1892.

178 *Vagus pneumonie* (Schou).

198 Varicosus conjunctivae (Gombert), Rech. Exper. Microbes Conjunctives, Paris, 1889.

59 Venenosus (Vaughan) Am. Jour. Med. Sci., 1892, 107.

" brevis (Vaughan), l. c.

" invisibilis (Vaughan), l. c.

193 Vermiculosus (Zimmermann), M.

76 *Vesiculosum.*

245 Vignal. B. lg. of : A. Ph. 86.

155 *Vignals B. j.* of.

290 Violaceus (Dyar).

293 Violaceus Lauerentius (Jordan), Mass. B. Health, 1890, 838.

290 " Sacchari (Ager-Dyar), Ager. N. Y. Med. Jour., 1894, 265, Dyar, l. c., 369.

220 *Virescens* (Frick), V. A., 116.

304 Viridis (Lesage), A Ph. 88 ; Cattrèlineau, A. P. 1806, 4.

314 " *pallescens* (Frick), V. A., 116,

34–327 Viridescens non-liquefaciens (Ravenel), l. c., 14.

313 *Viscosus* (Frankland), Z. VI.

72 " Cervisiae (Van Laer), Kramer, M. II, 119.

144 " lactis (Adametz), Landw. Jahrb. 01 ; Kramer M. II, 26.

205 Vulgaris (Proteus) (Hauser), Ueber Fäulnissbakterien.

Leipzig, 1885 ; Levy. Arch. Exp. Path. XIV, heft, 5-6.

Z. H. XII, 525 ; Tito Carbone C. VIII, 768.

Booker C.X., 284 ; Schnitzler, C. VIII, 793, XIV, 218.

Hauser XII, 630 ; Escherich. Darm. Bak.

Sauglings, Stuttgart, 1886.

Sanfelice. Atti. Accad. Med. Rom. 1890.

89 *Wildseuche* (Hueppe).

353 Zea (Burrill), Ills. Expt. Sta., 1889.

217 Zenkeri (Proteus), (Hauser), see Vulgaris, Kuhn, A. H. XIII., Petri A. K. G., VI. I.

277 Zeta (Dyar), l. c. 369.

216 Zopfi (Proteus), Kurth. B. Z. 83 Escherich M. M. W., 1886, 1.

102 Zurnianum (List), Adametz, Bak. Nutz—Trinker, Vienna, 1888, Dyar, l. c. 362.